PRACTICAL VULNERABILITY MANAGEMENT

PRACTICAL VULNERABILITY MANAGEMENT

A Strategic Approach to Managing Cyber Risk

by Andrew Magnusson

**no starch
press**

San Francisco

PRACTICAL VULNERABILITY MANAGEMENT. Copyright © 2020 by Andrew Magnusson.

Printed in USA

First printing

24 23 22 21 20   1 2 3 4 5 6 7 8 9

ISBN-13: 978-1-59327-988-2 (print)
ISBN-13: 978-1-59327-989-9 (ebook)

Publisher: William Pollock
Executive Editor: Barbara Yien
Production Editors: Katrina Taylor and Janelle Ludowise
Cover Illustration: Gina Redman
Interior Design: Octopod Studios
Developmental Editors: Alex Freed and Athabasca Witschi
Technical Reviewer: Daniel E. Dumond
Copyeditor: Anne Marie Walker
Compositor: Kim Scott, Bumpy Design
Proofreader: Paula L. Fleming

For information on distribution, translations, or bulk sales, please contact No Starch Press, Inc. directly:
No Starch Press, Inc.
245 8th Street, San Francisco, CA 94103
phone: 1.415.863.9900; info@nostarch.com
www.nostarch.com

Library of Congress Cataloging-in-Publication Data

Names: Magnusson, Andrew, author.
Title: Practical vulnerability management : a strategic approach to managing cyber risk/ Andrew
Magnusson.
Description: San Francisco, CA : No Starch Press, Inc., [2020] | Includes
    index. | Summary: "A hands-on guide to improving an organization's
    computer security and developing scanning tools on a budget. It starts
    by discussing the components of a vulnerability management program and
    then shows the reader how to build a free or low-cost system to
    automatically handle the repetitive aspects of vulnerability management"
    -- Provided by publisher.
Identifiers: LCCN 2020016647 (print) | LCCN 2020016648 (ebook) | ISBN
    9781593279882 (paperback) | ISBN 9781593279899 (ebook)
Subjects: LCSH: Computer networks--Security measures.
Classification: LCC TK5105.59 .M36167 2020 (print) | LCC TK5105.59
    (ebook) | DDC 658.4/78--dc23
LC record available at https://lccn.loc.gov/2020016647
LC ebook record available at https://lccn.loc.gov/2020016648

To Jessica

About the Author

Andrew Magnusson fell sideways into the information security field almost 20 years ago and never looked back. He started off as a firewall administrator, and has since tried his hand at security engineering, vulnerability analysis, and consulting. He presently heads up the customer engineering team for strongDM. He lives in Rhode Island with his wife, daughter, and two cats.

About the Tech Reviewer

A loving husband and father of twins, Daniel E. Dumond is a seasoned security practitioner and business leader with over 21 years of experience. Throughout his career, he has held many technical and senior leadership roles in the information security space, and traveled the globe to help build and deliver innovative security products and services for the world's most important organizations. Dan has a passion for programming and automation, which he uses to tackle some of the most daunting security challenges.

BRIEF CONTENTS

CONTENTS IN DETAIL

4

AUTOMATING VULNERABILITY MANAGEMENT   **25**

5

DEALING WITH VULNERABILITIES   **31**

6

ORGANIZATIONAL SUPPORT AND OFFICE POLITICS   **37**

PART II: HANDS-ON VULNERABILITY MANAGEMENT   **45**

7

SETTING UP YOUR ENVIRONMENT   **47**

10
MAINTAINING THE DATABASE

97

11
GENERATING ASSET AND VULNERABILITY REPORTS

103

12
AUTOMATING SCANS AND REPORTING

115

13
ADVANCED REPORTING

123

14
ADVANCED TOPICS   139

15
CONCLUSION   155

INDEX   163

ACKNOWLEDGMENTS

This project was by no means a one-man job. Many others were of great help to me in conceiving, writing, testing, and finally, publishing this book.

First, I want to thank my parents, Jan Slaughter and Phil Magnusson. Without their love and support, raising me to love words and to understand and harness computers, this book would not exist.

My wife, Jessica McKay-Dasent, was a constant and unconditional support during the time I wrote and revised this book. From the time I conceived it to the time it was completed, we got married, bought a house, and had our daughter Artemis. She was an amazing and supportive partner throughout, even as I slipped away for hours at a time to write and rewrite.

The editorial and support staff at No Starch Press, including but by no means limited to, Zach Lebowski, Alex Freed, Annie Choi, Barbara Yien, Janelle Ludowise, Katrina Taylor, Bill Pollock, and Athabasca Witschi, helped shape this book into its current state. They contributed innumerable improvements, large and small, to the raw material I provided them.

Dan Dumond performed an excellent technical review, making the text and scripts cleaner and more resilient in the process. Any remaining errors in the book and in the code are mine alone.

Annie Searle, a teacher and friend, was instrumental in instilling in me a risk-management mindset, and was kind enough to review early drafts of several chapters of the present work.

While I was at Mandiant Consulting, several colleagues generously supported this undertaking: Elliott Dorham, Mike Shingler, Dennis Hanzlik, and Jurgen Kutscher.

Finally, I'd like to thank my colleagues at strongDM, particularly Justin McCarthy, Elizabeth Zalman, and Schuyler Brown.

INTRODUCTION

It's human nature to pay attention to the problems that are big and flashy, attracting lots of interest, such as *advanced persistent threat (APT)* groups—state-sponsored attackers. APT-linked attackers have compromised major retailers, financial institutions, and even government networks. But when we focus all of our attention on APTs and other headline-generating activity, we miss basic issues. Even though you have new firewalls protecting your system and powerful traffic-monitoring devices, if you don't keep up with the bread and butter of your security responsibilities, you're leaving many chinks in your system's armor. Neglecting the basics, like keeping your systems updated, can lead to serious consequences.

Consider this example: suppose you're an information security manager at a medium-sized e-commerce business. You've set up firewalls to block incoming traffic except for traffic to internet-facing services on systems in your *demilitarized zone (DMZ)*. You've turned on egress filtering to block unauthorized exit traffic. An antivirus is on the endpoints, and you've hardened your servers. You believe your system is safe.

But an old web service is running on an outdated version of Tomcat on a Linux server in the DMZ. It's a relic from an ill-advised foray into selling some of your company's valuable proprietary data to selected business partners. The initiative failed, but because you made some sales, you had a contractual obligation to keep that server up for another year. At the end of the year, the project was quietly shuttered, but the server is still running. Everyone has forgotten about it. But someone on the outside notices it. An attack comes in from a compromised server in Moldova, and your unpatched Tomcat server is vulnerable to a five-year-old Java issue. Now the attacker has a foothold in your network, and all your protections couldn't stop it. Where did you fail?

This guide demonstrates the value of strong information security fundamentals. These are the most important components of a successful information security program. Unfortunately, they're regularly neglected in favor of sexier topics, such as traffic analysis and automated malware sandboxing. Don't get me wrong; these are great advances in the state of the art of information security. But without a strong grasp of the fundamentals, investment in more advanced tools and techniques is futile.

Who This Book Is For

This book is for security practitioners tasked with defending their organization on a small budget and looking for ways to replicate functionality from commercially available vulnerability management tools. If you're familiar with vulnerability management as a process, you'll have a head start. To build your own vulnerability management system, you should be familiar with Linux and database concepts and have some experience in a programming language like Python. The scripts in this book are written in Python, but you can functionally re-create them in whichever modern scripting or programming language you prefer.

Back to Basics

You can consider a number of security topics as foundational, such as authentication management, network design, and asset management. Although these elements might not be exciting or interesting for an analyst to work on, they're of critical importance.

Vulnerability management is one of the foundational concepts of information security. A perfectly written and configured software package doesn't exist. Bugs are an inevitable part of software, and many bugs have security implications. Dealing with these software vulnerabilities is a perennial issue in information security; the practice of vulnerability management is required for a baseline level of security that can serve as a trusted foundation upon which to deploy more advanced and specialized tools.

Vulnerabilities affect an organization's IT infrastructure at all levels, so vulnerability management affects all aspects of an IT security program. Endpoint security relies on workstations and servers being up-to-date with the latest software versions to minimize the attack surface. Zero-day vulnerabilities are always a concern. But removing the low-hanging fruit of known (and sometimes long-standing) vulnerabilities makes it more difficult for attackers to compromise an endpoint and gain a foothold in your environment. Network security does its best to ensure that only necessary traffic passes among internal network segments and to and from the internet. But if systems or network devices contain known vulnerabilities, even otherwise legitimate traffic might contain network-based attacks using known and trusted protocols. *Identity and access management (IAM)* restricts users to the specific systems and data to which they're entitled. But if the identity systems are vulnerable, attackers can simply sidestep them.

If your environment has a baseline level of security, any countermeasures you put in place can't be easily bypassed by exploiting known vulnerabilities. Let's consider an analogy: after World War I, France tried to protect itself from Germany by building a long line of forts and entrenchments along its German border. It was named the Maginot Line after the French minister of war. But when World War II began, the Germans ignored the barrier by simply going around it, invading France across the Belgian border instead. All of that expensive defensive infrastructure was irrelevant. The same goes for your environment. If it doesn't have a foundational level of security, any additional countermeasures are no more than a Maginot Line. Attackers can easily avoid them because there is an easier path elsewhere. But by establishing a vulnerability management baseline and maintaining it via an active vulnerability management program, you can trust that additional security measures will add real value to your security program.

Vulnerability Management Is Not Patch Management

Patch management, perhaps in conjunction with a full *software configuration management (SCM)* system, keeps track of the versions and patch levels of servers and endpoints across an enterprise. It can push patches remotely to keep systems up-to-date. But although traditional patch management and vulnerability management (as described in this guide) share many similarities, the underlying assumptions are very different.

Patch management assumes that patches are available, a patch management system can manage all the devices on the network that need patches, and there is enough time and manpower to apply all patches. But in real environments, it's very rare for all of these conditions to hold. Devices exist that aren't managed by the SCM: for example, network devices like routers and firewalls, test machines, abandoned servers, and devices running operating systems that aren't compatible with SCM agents. All these components are invisible to a typical SCM deployment and could easily become out-of-date without anyone noticing. Even if automated patching is practicable for endpoints, often you must handle servers and network devices manually, because automatically patching a server might lead to downtime when the organization can least afford it. On the other hand, manually patching servers and network devices takes time that overworked IT staff often can't spare.

Vulnerability management takes a more pragmatic approach. Instead of asking, "How can we apply all of these patches?" vulnerability management asks, "Given our limited resources, how can we best improve our security posture by addressing the most important vulnerabilities?" Vulnerability management looks at the problem through a risk management lens. We start with the full domain of vulnerabilities that exist on networked devices—managed and unmanaged—and determine which of these vulnerabilities present the highest risk to the organization's security. Once we've gathered that data, we have enough information to prioritize patching and remediation activities. If after this process is complete we have the capacity to apply more updates and remediation, so much the better. But by looking at the highest-risk issues first and using our limited time and resources wisely, we can improve the system's security posture significantly with comparatively little effort.

Main Topics Covered

This technical guide is divided into two main parts: conceptual and practical. In the first part, you'll learn about the concepts and components of the vulnerability management process. In the second and larger part, you'll look at a practical approach to building a free or low-cost vulnerability management system. Although you can follow the guide exactly, it's most important for you to understand the concepts behind each script to adapt it to your own needs. Toward the end of the book, you'll explore topics you might want to tackle once your vulnerability management system is up and running. One of those topics is purchasing a commercial tool to improve your vulnerability management program when you have the budget to do so.

How This Book Is Organized

Although there's a natural flow from chapter to chapter and part to part, from theoretical to practical guidance, if you're an experienced practitioner, you can jump to the specific topics of most interest. Similarly, the scripts naturally build from one to the next. But you can apply them on a piecemeal basis, depending on which tools and processes are already in place in your environment.

A summary of each chapter follows:

Chapter 1: Basic Concepts introduces the fundamental ideas of vulnerability management and its connection to risk management.

Chapter 2: Sources of Information discusses the various types of data you'll need to collect to conduct the vulnerability management process.

Chapter 3: Vulnerability Scanners explores the process of scanning the systems in your network to find vulnerabilities.

Chapter 4: Automating Vulnerability Management explains how to build an automated system to collect and analyze the data you collect.

Chapter 5: Dealing with Vulnerabilities describes what to do about the vulnerability information you gather: patch, mitigate, or accept the risk.

Chapter 6: Organizational Support and Office Politics provides information on how to accomplish vulnerability management in your organization.

Chapter 7: Setting Up Your Environment explains how to put together the underlying OS, install required packages, and write a script to keep everything up-to-date.

Chapter 8: Using the Data Collection Tools discusses how to use Nmap, cve-search, OpenVAS, and Metasploit.

Chapter 9: Creating an Asset and Vulnerability Database shows you how to import scan results into the database.

Chapter 10: Maintaining the Database covers adding keys and culling old data.

Chapter 11: Generating Asset and Vulnerability Reports delves into creating basic CSV reports for assets and vulnerabilities.

Chapter 12: Automating Scans and Reporting describes writing a script to automate Nmap and OpenVAS scanning and periodically generate reports.

Chapter 13: Advanced Reporting discusses advanced reports using HTML.

Chapter 14: Advanced Topics explores creating an API, considering automatic exploitation, and entering the cloud.

Chapter 15: Conclusion wraps up the book by providing information on future security trends and how they might change your vulnerability management process.

Outcomes

This book's goal is to take you from having no vulnerability management knowledge to having a functional vulnerability management program so you can generate accurate and usable vulnerability intelligence. This intelligence can help you increase your understanding of your organization's vulnerability landscape and improve the organization's overall security posture. By working through this guide, you'll strengthen your organization's vulnerability management capabilities, which is one of the fundamentals of a successful information security program.

Get the Code

As you are working through the steps to build yourself a vulnerability management system, you can always check the GitHub repository at *https://github.com/magnua/practicalvm/*. This repository contains all of the code in this book, as well as a few example configuration files that you can use in your own environment. Pull requests and suggestions are welcome!

Important Disclaimer

As is the case with most computer security tools or practices, you can use the tools and techniques in this guide offensively as well as defensively. Scanning can be an adversarial and malicious activity, and you should only perform it on systems (and networks) that you own or those you've been authorized to scan. I repeat: *do not scan or otherwise probe systems that aren't yours*. Even when used properly, such tools can potentially cause negative outcomes including, in extreme cases, system crash and data loss. Be aware of the potential risks before engaging in any scanning or exploitation-related activity.

PART I

VULNERABILITY MANAGEMENT BASICS

1

BASIC CONCEPTS

Before you dive into vulnerability management, you should first understand some basic information about vulnerabilities. You might already be familiar with vulnerabilities and their varying risk levels. If so, consider this chapter a refresher to prepare you for the more advanced topics to come. This chapter isn't an exhaustive primer of information security concepts, but it should be enough to ensure that the rest of the book is comprehensible.

The CIA Triad and Vulnerabilities

The three main pillars of information security are *confidentiality* of information (who can access data), *integrity* of information (who can modify data), and *availability* of information (whether data is available to authorized users). These three factors are known as the *CIA triad*. Although it isn't a perfect model, the terms aid in discussing and categorizing security vulnerabilities.

Software, firmware, and hardware have bugs, and although not all bugs are serious, many have security implications. If you can enter improper input into a program and cause it to crash, not only is that a bug, it's a *vulnerability*. But when you enter improper input and all it does is change the onscreen text color, presuming the text is still visible, that bug isn't a vulnerability. Well, it isn't until someone clever figures out how to leverage that bug to cause security-related issues. In short, a vulnerability is a weakness in an information system that an attacker can leverage in a way that has security implications. Typically vulnerabilities are due to bugs, but these weaknesses could stem from flaws in the code logic, poor software design, or implementation choices.

Because a bug must have implications for the confidentiality, integrity, or availability of data—or an entire information system—to be considered a vulnerability, the major vulnerability types map directly to the CIA triad. Denial-of-service (DoS) vulnerabilities impact the availability of data: if authorized users can't access the system, they can't access the data either. Information disclosure vulnerabilities impact data confidentiality: they permit unauthorized users to access data that they couldn't otherwise access. Similarly, information modification vulnerabilities allow unauthorized users to modify data, so these vulnerabilities impact data integrity.

A fourth vulnerability category involves code execution and command execution. These vulnerabilities allow attackers to execute specific commands or arbitrary code on a system. The attacker has either limited or complete access to the system, depending on the user level at which this code executes, and can affect all three portions of the CIA triad. If an attacker can run commands, that person might be able to read or modify sensitive data or even shut down or reboot the system. Vulnerabilities in this category are the most severe.

Some vulnerabilities might fit into more than one category, and the categorization (and severity) could change as attackers begin to better understand the vulnerability and exploit it more thoroughly. Because the vulnerability landscape changes constantly, you need an effective vulnerability management program to keep abreast of developments.

What Is Vulnerability Management?

Vulnerability management is the practice of staying aware of known vulnerabilities in an environment and then resolving or mitigating these vulnerabilities to improve the environment's overall security posture. Although this definition sounds simple, it entails a number of interdependent activities. I'll discuss each of these activities in more detail in the following chapters. For now, let's look at the vulnerability management life cycle's major components (see Figure 1-1).

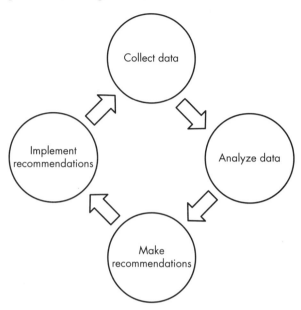

Figure 1-1: The vulnerability management life cycle

The first step is to understand the current vulnerability environment. To do so, you need to collect data about your systems to determine the vulnerabilities that exist on them. The next step is to analyze that collected data as well as security-related data from other sources.

Your data analysis results will help you make recommendations about the actions needed to improve your security posture. These recommendations might include installing patches or applying mitigations, such as firewall rules or system-hardening techniques. The next step is to implement recommendations. Once this is complete, the cycle begins again: you collect another round of systems data and the vulnerabilities that remain after analysis and mitigation, as well as new vulnerabilities that weren't apparent in the previous cycle.

The management process is neither short nor simple. Finding vulnerabilities can be easy, but dealing with them and improving your security baseline will be ongoing. The process will also involve many different roles and business processes throughout the organization.

Let's look at each step in more detail.

Collecting Data

You can split the collection component into two major categories: internal and external data collection. We'll look at each in turn.

Internal data collection involves gathering information about your organizational environment. This data includes information about the hosts on your network—endpoints and network devices—and vulnerability information about each host. Host information can come from an exploratory scan using a network-mapping tool (like Nmap), an asset database tool, or a *configuration management database (CMDB)*. If you have only a spreadsheet that contains data about your servers and workstations, it won't be sufficient. For vulnerability management to be successful, you need to start with accurate and complete data. A spreadsheet you create and update manually won't reflect the actual hosts and network information that live in your environment.

Vulnerability data comes from one source: *vulnerability scanners.* These tools discover vulnerabilities by interacting with devices, either through network-based scans or host-based agents. Network scanners reach out to every IP address within a range, or a specific list of IPs, to determine which ports are open, which services are running on those ports, the operating system (OS) versions and relevant configurations, and software packages running on each device. Host-based scanless agents query the system directly to determine running services and version information. Both approaches have benefits and drawbacks, which I'll discuss in more detail in Chapter 3.

The internal data you collect quickly becomes stale—this is especially true of vulnerability information—so you must gather it regularly. Even though you might not add or remove hosts frequently, vulnerability information changes daily: people install new software packages or perform updates, and new vulnerabilities are discovered and publicly disclosed. Regular scanning and routine scanner updates to incorporate new vulnerability information ensure that you have accurate and complete data about your current environment. On the downside, regular scanning might have negative effects. But you must balance this risk against the importance of having accurate vulnerability data. I'll discuss this trade-off in Chapter 2.

Information like network configurations and other advanced data sources, although potentially useful in your analyses, are outside the scope of this guide. But the same warning applies: if the information isn't recent and thorough, your entire analysis is less useful to you. Fresh data is good data.

External data collection encompasses the data sources that come from outside your organization. This information includes public vulnerability details, embodied by the constantly growing mass of *common vulnerabilities and exposures (CVE)* data that NIST (the *National Institute of Standards and Technology*) provides; public exploit information from the Exploit Database and Metasploit; additional vulnerability, mitigation, and exploit detail from open sources like CVE Details (*https://cvedetails.com/*); and any number of proprietary data sources, such as threat intelligence feeds.

Although this information comes from outside your organization, you can still remain up-to-date at all times by either querying online sources directly or keeping local data repositories. Unlike local data collection, which might cause issues in your environment, collecting data from third-party sources is as easy as reaching out and getting it. So, you have no reason—except perhaps to save data transfer costs—not to update these sources daily or even keep a live connection in the case of threat intelligence feeds.

Analyzing Data

Once you've collected internal and external data, you need to analyze this data to gain useful vulnerability intelligence about your environment.

Vulnerability information alone, as anyone familiar with a scanner report can tell you, is overwhelming for any environment larger than a few devices. Scanners will find many vulnerabilities on nearly every device, and separating important vulnerabilities from the unimportant ones can be difficult. Worse, if all you have is a thousand-page scanner report, you'll have a hard time deciding which remediation tasks to assign to an already over-worked systems administrator.

You can approach this problem in two ways. One way is to try to reduce the list of vulnerabilities to a more manageable length, known as *culling*. The other is to try to rank the vulnerabilities in order of importance, known as *ranking*.

Culling is straightforward: it's a binary yes-or-no decision you make on every vulnerability. The criterion for accepting a vulnerability might be, for example, the vulnerability is newer than a certain date, there are known exploits, or it's remotely exploitable. You could also combine any number of these binary filters to cull the list even further. Only if a vulnerability matches the criteria would you take the time to analyze it further.

Ranking requires a criterion using some sort of scale. For instance, you could rank a set of vulnerabilities based on their effects on confidentiality, integrity, or availability. Or, you could use the *Common Vulnerability Scoring System (CVSS)*, which is a 1-to-10 scale that takes into account a vulnerability's severity along all three of the CIA triad's axes. If you have a strong understanding of your organization's risk landscape, you might have your own scoring system that focuses on internally developed risk metrics.

Although these two methodologies have different focuses, you can convert between them. You can use a binary categorization, such as exploitability, to rank rather than to cull, resulting in a list that is split into two groups. In contrast, you can use a ranking metric to cull by setting a threshold. For example, you could set a culling threshold of a CVE score of 5 and ignore any vulnerability with a lower score. Given a metric for categorizing vulnerabilities, you should then decide whether you want to use this category as a ranking or as a culling metric, or both.

Because culling results in a smaller dataset to analyze, whereas ranking is an analysis method in itself, consider using both. By first culling the vulnerability set, you can limit your subsequent analysis to vulnerabilities that you must address, which makes analysis faster and more relevant. Once you identify the most critical vulnerabilities, you can rank the remaining vulnerabilities to more easily determine their relative significance.

In this book's scripts, I use a simple cull-rank profile, which you can modify or replace based on your organization's needs. This profile uses the CVSS score and exploitability as metrics (see Figure 1-2).

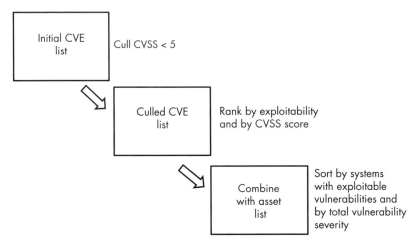

Figure 1-2: A simple cull-rank profile to filter important vulnerabilities

You first cull vulnerabilities with a low CVSS score because they're not severe enough to analyze further. Next, you rank the remaining vulnerabilities by exploitability and then by CVSS score, from high to low. You combine this list with the asset list. Then rank the resulting list first by the number of exploitable vulnerabilities per system and then by the total severity of vulnerabilities found on the system. The resulting list shows the systems with the highest risk at the top.

Applying Cull-Rank to a Real-World Example

Let's look at an example of how the cull-rank analysis process might work in a real-world scenario: let's say you just ran a vulnerability scan against your main end-user network segment—a Class C network with 256 total addresses of which 254 are usable. You know the segment includes numerous Windows hosts as well as a handful of printers and miscellaneous devices. The scan result shows a list of approximately 2,000 total vulnerabilities spread across 84 devices.

You work through the list and cull vulnerabilities with a CVSS score less than 5, cutting your list to about 500 vulnerabilities on 63 devices. At this point, you have only 38 unique vulnerabilities—most of the vulnerabilities exist on multiple hosts—which means you only need to look at each of those 38 vulnerabilities once. By this measure, you've already cut the list of items to investigate by about 92 percent. To determine which of the remaining vulnerabilities you need to investigate, you'll apply several rankings.

First, find out whether any of these 38 unique vulnerabilities have publicly known exploits. If they do, you need to address those vulnerabilities first. Second, establish what the CVSS severity of each vulnerability is. Higher severity means greater consequences of compromise, so you should focus on the more severe vulnerabilities.

Before you execute the third ranking methodology, look at what you have so far. Of your 38 unique vulnerabilities, 3 have known exploits, and the remaining 35 have been sorted in order of CVSS severity.

Now you can apply the final ranking: combine the list of vulnerabilities with the actual vulnerable hosts. For each host, determine how many vulnerabilities it has and the severity of those vulnerabilities. Once you've done this, you'll have a clear picture of where you need to focus your remediation efforts.

In this example, among those 63 hosts with vulnerabilities, 48 have one to two vulnerabilities of severity no higher than 7, whereas 11 have up to 15 vulnerabilities with one or two in the critical range (CVSS of 9 and higher). The last four contain all the rest of those 500 total vulnerabilities among them—an average of 125 on each host, including all three exploitable vulnerabilities! Clearly these systems need heavy remediation, and you have a good argument for addressing the situation immediately.

Making Recommendations

Now that you have a list of hosts and vulnerabilities that is sorted by risk to your organization, the next step is to recommend actions to remediate the vulnerabilities. You'll start with the highest risk and work your way down the list. If you're working in a small environment, you might be responsible for this step; in a larger organization, this step might consist of a longer process that involves working with system and application owners as well as other stakeholders.

The two major types of remediation are *patching* and *mitigation*. Patching is simple: you apply the patch that resolves the vulnerability in question. Mitigation is more complex and is context dependent.

If a patch isn't available or if it's infeasible to apply one, you need to look at other ways to address the risk. Perhaps changing a configuration will prevent a specific vulnerability from being exploited. Perhaps the vulnerable service isn't needed outside specific IP ranges so you can protect it with firewall rules or router *access control lists (ACLs)*, reducing the exposure. Perhaps an existing *intrusion detection system (IDS)* or *intrusion prevention system (IPS)* needs additional rules to detect whether someone is attempting to exploit that specific vulnerability and block it. All of these are examples of vulnerability mitigation, and the correct response will depend on your environment.

Implementing Recommendations

With recommendations in hand, you can finally approach the system and application owners to suggest they implement the proposed remediation actions. If they were involved in the recommendation process, this step should be straightforward. If the recommendations are unexpected, you'll need to explain the security risks and the reasons for the recommendations you've developed. I'll discuss this process in Chapter 6. At this stage, you should all agree on a timeframe for the implementation.

Once those responsible have implemented the recommendations—via patching or mitigation—the final step is to verify that the changes have been made and are effective. Because mitigating controls vary widely, determining that they're in place and effective is largely a manual process. But with patching, you can verify the changes by scanning again to see whether the vulnerabilities still exist. This returns you to the first phase—collecting data. The cycle starts over, and the new scans will validate remedial actions and discover new vulnerabilities.

Vulnerability Management and Risk Management

Vulnerability management is closely tied to the enterprise's risk management goals. This technical guide doesn't focus on information risk management as a whole. But it's important to understand where vulnerability management corresponds to risk management. Without a functional vulnerability management program, the enterprise's IT risk management goals will be difficult, if not impossible, to achieve.

The overall IT risk management framework is similar to vulnerability management. Generally, the IT risk management stages are to identify critical assets, identify and rank risks, identify controls, implement controls, and then monitor the controls' effectiveness. Risk management is also a continual process rather than a one-time event with a defined endpoint. So where does vulnerability management fit into this process?

Different phases of vulnerability management map to different phases of the risk management process (see Table 1-1). For instance, identifying assets in the risk management framework is directly related to collecting asset and vulnerability data.

Table 1-1: Mapping Vulnerability Management to IT Risk Management

Vulnerability management	IT risk management
Collect data	Identify critical assets
Analyze data	Identify and rank risks
Make recommendations	Identify controls
Implement recommendations	Implement controls
(Collect data)	Monitor controls

But these mappings are only part of the process. Vulnerability-related risks discovered through the vulnerability management process might lead an organization to consider controls that don't directly resolve the vulnerabilities, such as implementing a protocol-aware firewall. Although a measure like that would be effective against certain exploits, it would also mitigate various other risk types. In addition, regular vulnerability management data collection is useful not only for identifying assets and risks but also for monitoring the controls' effectiveness. For example, you implement a firewall as a control, but the next scan indicates that it's misconfigured and not filtering the traffic it's intended to block.

Because this guide isn't an information risk management cookbook, we'll leave this discussion here and continue to an in-depth exploration of vulnerability management. But if you're interested in understanding information risk management methodology and procedures, I recommend looking into *NIST 800-53* and *ISO/IEC 27003*, *ISO/IEC 27004*, and *ISO/IEC 27005*. You can find each with a Google search.

Summary

This chapter provided you with a crash course in vulnerability management and its place in the larger IT risk management framework. You learned about the general vulnerability management process that you'll follow throughout the remainder of this book and previewed the steps to take once you have actionable vulnerability intelligence.

In the next chapter, we'll look more closely at the vulnerability management process and get a step closer to implementing your own vulnerability management system.

2

SOURCES OF INFORMATION

To have a successful vulnerability management program, you need information from several data sources. This chapter introduces you to each of these sources. In the next chapter, you'll see how they all come together to give you a useful vulnerability landscape for your organization.

Asset Information

Despite the importance of asset information, many organizations, large and small, don't have a full—or even fragmentary—understanding of what is on their networks. Perhaps you use a spreadsheet that you pass among network administrators and update intermittently. Or maybe you have a database of Windows desktops using a CMDB or endpoint management product. But to perform vulnerability management, you need a complete inventory of IP-connected devices and any additional data that you can glean about each host. Non-networked devices, although important to an overall risk assessment, are outside the scope of an automated vulnerability management program.

Obtaining a list of hosts—and even a wealth of additional information—is straightforward. You can use a network-scanning tool, like Nmap, or a vulnerability scanner, like Nessus or Qualys (which you'll need anyway to collect vulnerability data), to do a network sweep and find live hosts. But these scans can be obtrusive and might cause application or even OS crashes. So, you need to carefully plan for an information-gathering scan.

New devices are added to networks all the time, and although most organizations have a change management policy in place, this is no guarantee that changes aren't made without following policy. To have updated and trustworthy asset information, you must perform discovery scans on a regular basis across the entire network.

Ideally, you would run these scans on a schedule. But the risks of regular scanning (which you'll learn about in the next section) might mean that your organization isn't comfortable having these scans done without a human monitoring them, ready to stop the scans if any issues crop up. If this is the case, you'll need to run fewer discovery scans, on a manual basis, and import the resulting data into your datastore.

CHANGE MANAGEMENT

Any organization with risk management in place will have a change management system to ensure that systems and networks remain in a stable state and to document any changes to that state. These systems can range from an email chain for change requests, approvals, and coordination to a full commercial change management system encompassing ticketing and configuration management.

Change control alone must not, and cannot, be the only control in place for IT changes. There are always ways around it. Administrators apply—or fail to apply—patches, add and change network routes to troubleshoot issues, buy and connect new network devices, or enable new services to fulfill a perceived business need without creating the necessary change control paper trail. Thus, you can't trust what a change management system says about the IT infrastructure's state.

Vulnerability Information

Once you have a complete account of all devices on the network, you'll configure a vulnerability scanner to do a deep scan of each device and discover any known host vulnerabilities. For example, a scanner might determine that a Windows server is running a version of the IIS web server that is vulnerable to a directory traversal attack: the consequence is possible information disclosure.

When configuring and scheduling scans, carefully look at the available scanner options and tailor the settings to your environment and risk tolerance. The same goes for scheduling the times and scopes of scans. For example, you can scan some of your network sections, such as endpoint segments, every day. The reasons are that the risk of downtime is limited and the consequences aren't severe if a user's workstation is briefly offline. But scanning your critical systems, such as core production databases, might be too risky to do outside of scheduled maintenance windows. You need to understand the trade-off between getting fresh data and risking downtime.

By their very nature, network vulnerability scanners will only find vulnerabilities that are discoverable over a network connection. If a locally exploitable vulnerability is in a desktop application on a Windows endpoint, a network scan won't find it. For example, a network scanner won't find CVE-2018-0862—a vulnerability in Microsoft Equation Editor that an attacker can only exploit by opening a crafted Word or WordPad document. The reason is that Microsoft Office applications in general aren't detectable via a network scan.

To plug this hole, you could use an endpoint scanner (for example, the Qualys "scanless agent") or a *software configuration management (SCM)* tool or CMDB to gather a list of deployed software versions and determine known vulnerabilities by checking against a vulnerability database. Despite these limitations, having an accurate account of just network-discoverable vulnerabilities is an excellent start.

I'll cover vulnerability scanners in more detail in Chapter 3.

Exploit Data

Although a lot of information is available on a per-vulnerability basis, you can do more by combining data sources. The lowest-hanging fruit is exploit data. Information about publicly available exploits is widely accessible and often searchable. For example, the Exploit Database website (*https://www.exploit-db.com/*) has a searchable index of public exploits. Also, Metasploit, which I'll discuss in Chapter 14, has a large archive of usable exploits and a command line tool to easily deploy these exploits against target systems. Most exploits are associated with a particular vulnerability—a specific CVE ID. You can use the CVE ID to correlate exploit information with vulnerability information that you already possess.

Addressing an exploitable vulnerability is likely a higher priority to your organization than a vulnerability that isn't yet known to be exploitable. But not all exploits are equal. For instance, an exploit that enables arbitrary code execution is more severe than one that causes DoS or even one that permits reading arbitrary data. Knowing the consequences of an exploit is very useful for prioritizing exploits with more granularity.

Advanced Data Sources

The following list contains a few specialized and advanced data sources. Although largely outside the scope of this book, they're valuable references.

Threat intelligence feeds   These feeds include information about the current threat landscape: threat actors and groups, the exploits currently being used in exploit kits, and the vulnerabilities with privately available exploits that aren't yet public knowledge. Use this information to determine which vulnerabilities are currently a higher risk to your organization. Because these threat feeds contain fresh data, you should use the feed data as soon as it comes in to get a timely assessment of your exposure to newly discovered threats. Numerous free and paid threat feeds are available, such as iSight Threat Intelligence, iDefense Threat Intelligence, and industry-specific threat feeds, like the one provided by FS-ISAC.

Proprietary exploits   Although it's expensive, adding proprietary exploit data (sometimes known as exploit kits) to the publicly available information from Exploit Database and Metasploit broadens the range of exploits that you can match against your vulnerability data. Sources range from commercial threat intelligence sources that commission their own exploit research to decidedly gray- or black-market options, such as independent researchers selling newly discovered vulnerability and exploit information to the highest bidder. Whatever the source, proprietary exploit information will help you better prioritize your own vulnerability data based on exploits you would otherwise be unaware of.

Network configurations Use network configurations from routing devices like routers, firewalls, and managed switches to create a model of your network. By combining this information (which subnets route to which, which ports are accessible from where) with vulnerability and exploit data, you get a deep understanding of your network attack surface. For example, if a Tomcat exploit exists for an internal web application server but your router configuration indicates that this server is accessible only to a limited list of source IP addresses, it might be of less concern to you than if it were accessible to the internet at large. You might already have network configuration information, especially if you have a centralized configuration repository, such as SolarWinds. On the downside, it takes significant work to integrate this data with your existing vulnerability data. Some commercial vulnerability management products contain built-in functionality to ingest network configurations.

Summary

Each of the data sources discussed in this chapter contributes an important set of data to your vulnerability management system. Table 2-1 breaks down the data you can glean from each of these sources.

Table 2-1: Data Sources for Vulnerability Management

Data source	Important data
Host/port scanner (Nmap)	IP address MAC address Hostname Open ports (TCP and UDP) Service and OS fingerprinting
Network vulnerability scanner	(Same as above) Additional service fingerprinting and version detection Network vulnerabilities Local vulnerabilities (authenticated scans only)
Host-based vulnerability scanner	Local vulnerabilities
CMDB/SCM	OS details Deployed software details Configuration details Owner of the device Criticality of the device and application
Exploit databases	Exploit information Vulnerability mapping to exploits
Threat intelligence	Attacker and targeted industry intelligence Newly discovered, escalating, or widespread exploits
Exploit kits	Proprietary exploit information
Network configurations	Network topology and potential attack paths

In the next chapter, you'll take an in-depth look at vulnerability scanning.

3

VULNERABILITY SCANNERS

Although vulnerability management has several other components, the raw data you collect from your vulnerability scanner is the most important. If the scanner isn't configured correctly or if it's located in the wrong place, it won't give you the data you need for the rest of your vulnerability management process.

This book assumes use of a *network-based* scanner, which learns about a system by sending packets across the network and listening for particular responses. In this chapter, I'll discuss how network-based vulnerability scanners work and how to make the most of them in your environment.

What Vulnerability Scanners Do

A scanner discovers all it can about the OS and running services on every device you configure it to scan. Based on the discovered information, the scanner determines whether the device is susceptible to any known vulnerabilities. Once it finishes collecting a list of vulnerabilities on all the devices within the specified network ranges, the scanner produces a report. The report consists of a list of hosts, any information known about them, and which vulnerabilities exist on each host. You'll use this report as a major component of your vulnerability analysis.

How Vulnerability Scanners Work

Security administrators configure scanners to scan specific network ranges or individual systems, or *targets*. The scanner sends ping packets to all IP addresses in the given target—reaching out to hosts to determine what is active and responding. Once the scanner knows which IP addresses belong to live devices, it sends additional pings, connection requests to open ports, or packets crafted to elicit certain status messages or error responses. An administrator can configure the scanner to be more or less aggressive when probing. At higher levels of aggression, the scanner will send thousands of packets to each device to find out which ports are open and what kind of device it is.

Once the scanner recognizes what a device is and what services are running on it, it sends probes to determine additional information. For instance, if it detects that the device is listening on port 80 (a typical web server port), it will try to connect to the web server to identify which server software is running (and which version is in use). The scanner matches version information against its own internal vulnerability database. If, for instance, the device is running version 3.1 of a specific piece of software, and vulnerabilities in that software were fixed in version 3.2, the scanner reports that the device is vulnerable to those issues. In addition, some scanners have specific tests for certain vulnerabilities. These tests are useful in cases where other mitigations that prevent exploitation of that vulnerability have already been implemented.

Scanning isn't an exact science. Although fingerprinting often discovers the OS running on a device, exceptional cases might throw it off. For example, a customized network stack can make a machine appear to be running a different OS, or a rare OS without a good fingerprint available might be misidentified.

A similar level of uncertainty exists with vulnerability discovery. For instance, an HTTP server reports that it's running Apache 2.2.0, and the scanner infers that it's vulnerable to specific issues that weren't fixed until Apache 2.2.1. But that system's vendor backported those fixes to its customized version of Apache 2.2.0, so in fact the system isn't vulnerable. The scanner has no way of knowing this, so it reports the false positive that the system is vulnerable. Although you can minimize these errors, false positives are part of network vulnerability scanning.

How to Deploy Vulnerability Scanners

You have a number of choices to make about how you deploy scanners in your systems, including how to give them access to the networks they need to scan, what OS and hardware they run on, and how to configure them so they are effective in your environment.

Ensuring the Scanner Has Access

Scanner placement within the network is critical. If you're trying to scan any network segment other than the local network, the scanner will send packets through routers and maybe even firewalls. Those devices might have ACLs or firewall rules preventing certain sorts of traffic, both of which are likely to drop the scanner's probe packets. As such, you can use two general methods for deploying scanners.

First, you can open full access through any intervening network devices from the scanner to any network ranges you intend to scan. This might involve excluding the scanner's traffic from IPS policies. Opening full access ensures that packets aren't blocked between the scanner and its target, which could cause incorrect results.

Second, you can set up multiple scanners—each local, or close in the network topology, to the network segments to scan. For instance, if you want to scan a firewall *demilitarized zone (DMZ)*, place a scanner within the DMZ so it has direct access to the systems it's trying to scan. Both approaches have strengths and weaknesses.

RESTRICTED OR UNRESTRICTED SCANNING?

One school of thought holds that you should perform all scanning via a "normal" network environment. In other words, the scanner shouldn't be able to access anything that an unprivileged user on that network segment is blocked (by firewall rule or ACL) from accessing. An attacker won't have unfettered access, so why should your scanner? But you shouldn't be trying to emulate an attacker's point of view when you run scans. You want to get a complete view of the vulnerabilities across your enterprise. What if your attacker is in a different subnet? What if your attacker has already compromised a system in the "restricted" network? Your scanners need to see the entirety of the systems they're scanning to do their job.

Opening Full Access to Your Scanner

By configuring the network to accept and pass all scanner traffic and responses, you can locate the scanner anywhere in your network. You can also use a single scanner for multiple network segments, which reduces the cost of setting up a scanning environment. But opening full access from the scanner can also be dangerous; attackers can leverage holes in your router and firewall infrastructure. If an attacker compromises the scanner system, they could use its access to the rest of the network to compromise other systems more easily.

Setting Up Multiple Scanners

Setting up multiple scanners might seem to be the best approach. You don't need to open firewall ports or add ACLs. But this method has its own drawbacks.

First, it costs more to set up, because each scanner requires its own physical or virtual hardware. Also, if you're using a commercial scanner, the licensing fees could cost more than the underlying hardware.

Second, you might have a coordination issue: either you'll have to connect to each scanner directly to set up scans and retrieve the results separately, or you'll need some kind of hierarchical environment to manage the scanners from a central location. For example, Tenable sells Security Center to manage multiple Nessus scanners, and Qualys uses the cloud-based QualysGuard. This coordination adds time and costs to scanner deployment.

Additionally, you'll likely still need to open firewall ports and router configurations to ensure the analyst (or central control system) can access these scanners in various network locations.

Choosing Your OS and Hardware

Some scanners, such as Qualys, use their own appliances, so you don't have control over the underlying system. But others—including Nessus and OpenVAS—are applications that run on whichever platform you prefer. The OS you choose doesn't matter, as long as your scanner of choice supports it. You can use whichever platform you're most familiar with or whichever follows your organization's policy.

For the hardware, more power is always better. Scanners use a tremendous amount of RAM because they run a large number of concurrent tests against multiple targets, so it's best to beef up the RAM first. Generally speaking, two CPUs and 8GB of RAM should be enough for a small deployment. Less RAM will work, but your system might be unresponsive while the scanner is running. A fast connection (high speed with low latency) is also essential; otherwise, the network tests will take a long time and might even return false positives or negatives if they time out before completion.

Configuring Your Scanner

Once your scanner is set up and online, you need to tailor the scanner to your environment. Scanners have a long list of configuration options, often called *policies* or *templates*. They let you configure how fast your scanner sends out its packets, what types of tests to run, and many other choices.

The options ensure that the scanner returns useful results without overburdening your network or causing issues with the devices it scans. If you have a test environment, this is a good time to use it: configure your scan policies and then scan the test network. If you run into problems (for example, network congestion, or slowdowns, or device malfunctions caused by the scanner's actions), adjust your policy until you eliminate those issues. Once you're sure the scanner is properly customized, you can scan your live environment. Whether or not you can practice on a test network first, it's still best to run a few scans on a small portion of your live network before scanning the entire system. It's better to experience problems with a few systems—preferably those you manage that are geographically close—than to cause your entire production database environment to reboot!

Once you've tested your scan policy and are ready to scan larger parts of your network, or even the entire system, consider the best way to set up your scans' targets and schedules. There are a few reasons to break up your scans into more manageable pieces. If you run one enormous scan, it might take a very long time to complete—time you could otherwise be using to analyze scan data from smaller scans. If you have several scanners, you'll need to set up your scans so the correct scanners are targeting the correct networks. Additionally, the ideal times to run the scans might be different for different network segments. Although it might be fine to run scans on your workstation VLANs after hours, that might be prime processing time in some of your data center environments. Some sensitive or production-critical networks might simply be off limits until designated change windows, just in case something goes wrong and a system becomes unresponsive.

When you're planning your scans, inform other stakeholders of your scan plans and policies and involve them in determining appropriate scan targets and time windows. Involving others early in the planning process ensures you won't catch them by surprise if you run a scan that causes a domain controller to crash. You don't want the Windows administrators to successfully lobby for their systems to be exempt from future scans. To have a complete view of your network, you'll need buy-in, or at the very least grudging acceptance, of the scanning regimen before you begin. In Chapter 6, we'll talk more about how to get scanning and remediation accomplished in a business environment.

Some organizations simply won't allow regular automated vulnerability scans. In this case, manually run the scans with an analyst present to monitor the scan progress. Then, if any issues pop up on the scanned systems, the scapegoat—I mean, analyst—can halt the scan before downtime is incurred or prolonged.

Organizations with very stringent uptime requirements and extremely rare maintenance windows might use lab or redundant systems as scan targets instead of live systems. An additional set of servers—or an entire network—is duplicated, including the network configurations, OS, applications, and patch levels. Any scan data from the test systems will, in principle, be identical to any data from live systems.

But it can be very difficult to perfectly synchronize the patch levels of systems. Any difference in configuration can lead to differing scan results—for instance, if the host-based firewall on a lab system opens a different set of ports than on the live system. The bottom line is that there is no way to ensure accurate results without conducting scans against the real-world systems that you're tasked to protect.

Getting Results

Scanners can report their results in many ways: a plaintext file; a structured format like XML or CSV; or more readable formats like HTML, an RTF or Word file, or a PDF. Although the last options are preferable for reading the reports directly, we'll focus on the machine-readable output formats, such as XML and CSV. The reason is that running the scanner and collecting its results is only the first step. You'll still need to analyze this data to get useful vulnerability intelligence. Most, if not all, scanners can produce output in XML, and that is the format you'll use most in this guide. But any structured, computer-parseable format will do.

Summary

Vulnerability scanners can generate vast amounts of information, and it's important to understand what is useful to you as a vulnerability analyst. By properly deploying and configuring your vulnerability scanners, you can ensure that you're collecting only (or at least primarily) data that matters to you while disregarding irrelevant information. In some environments, it might make sense to deploy several scanners to get different views of your network or to gain visibility into otherwise heavily restricted network segments. You might also separate your scans into smaller targets to get results faster or to limit the chances of causing networkwide outages as a scanning side effect.

In addition, you must consider the broader operational environment you're part of. Scans can be intrusive and in extreme cases cause system downtime, so you need to carefully introduce regular vulnerability scanning. Involve other teams in the process of determining appropriate targets and schedules so you're not solely responsible if a scan goes wrong.

In the next chapter, you'll learn how to collect and analyze the data your scanners produce. You'll also be introduced to the automated methods for collection and analysis that you'll build in the practical portion of this book.

4

AUTOMATING VULNERABILITY MANAGEMENT

In this chapter, you'll learn how to programmatically compile your data sources to provide vulnerability prioritization and validation. As a result, you'll save time for more important work, such as improving your organization's security, rather than going cross-eyed staring at huge vulnerability data dumps.

Understanding the Automation Process

Automating your vulnerability management program consists of correlating information from the three main data sources—asset, vulnerability, and exploit information—as well as any additional accessible data sources. For a refresher on these data sources, refer to Chapter 2.

Information is correlated through two shared fields: one contains IP addresses that are shared between asset and vulnerability data. The other contains CVE/BID IDs (BID is short for Bugtraq ID) that are shared between vulnerability and exploit data. First, you use the IP address to correlate assets with vulnerabilities, and then you use the CVE ID to correlate vulnerabilities with exploits. The result is a useful database that provides a list of exploits per host, hosts per exploit, and more. Figure 4-1 shows this step-by-step process.

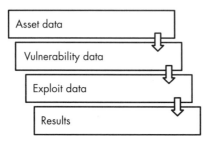

Figure 4-1: Correlating information to produce a useful database for vulnerability analysis

You break down each step, except for Results, into substeps: *collecting* the data and *correlating and analyzing* the data. You'll need to get all the data into one place before you can start analyzing it. In this book, you'll use MongoDB, a document-based database that excels in speedy queries over large volumes of data. But you can also accomplish this process through more traditional SQL databases by replacing the Mongo-specific code in the upcoming scripts with SQL connections and queries.

In each step of the process, you'll collect the relevant data, import it into your Mongo database, and perform appropriate analysis at that stage, before assimilating the next set of data. Once you've completed this process on your own, you'll find that certain levels of analysis are more useful to you than others. You'll then be able to streamline your process to highlight those analyses and downplay or set aside the rest.

Data Collection

In the first stage of the process, *asset data* analysis, you find assets, their network information, and the OS running on each asset.

Once you add in *vulnerability data*, you'll match vulnerabilities with specific assets and pinpoint hosts that are in the greatest need of vulnerability remediation. Important data points at this second stage include CVSS score, which describes the overall severity of the vulnerability; *attack vectors*—whether the exploit is local, remote, and so on; and specific *consequences of exploitation*, such as DoS or root code execution.

Next, you add *exploit data* to further prioritize among vulnerable hosts, highlighting hosts that are vulnerable to known exploits and hence at greater risk of exploitation by malicious actors. At each stage of the analysis process, you can generate reports with useful security-related information, which Table 4-1 summarizes.

Table 4-1: Data Sources and Their Potential Analyses

Data	Analysis
Asset data	Asset summary: a report on assets, their OS, open ports, and networking information
Vulnerability data	Vulnerability summary: discovered vulnerabilities on an asset or set of assets Vulnerability prioritization by CVSS, attack vectors, consequences: the same report as above but filtered to look for specific vulnerability types
Exploit data	Exploit matching and further vulnerability prioritization: a report focusing on exploitable vulnerabilities or those with certain exploitability characteristics

The two processes described in Chapter 1—culling and ranking—can take place at any of these stages, depending on the criteria you're using. For instance, an IP-based cull could take place as soon as you have asset data. On the other hand, prioritizing based on CVSS can't take place until you have vulnerability data.

By culling early, you can limit analysis work. But for simplicity of analysis, it's easiest to do the culling and ranking steps in one place, once you have all the relevant data. That way, if you want to change your analysis priorities, you can change your criteria in one place rather than looking in multiple scripts from different phases of the vulnerability management process.

Once you've combined the datasets and applied prioritization rules, you have a finished product: a list of hosts with relevant vulnerabilities per host, sorted with the highest risk hosts/vulnerabilities at the top.

Automating Scans and Updates

You can gather all the information discussed so far manually. For instance, you can run ad hoc Nmap and vulnerability scans and manually look up information about known exploits. But you experience the real power of a vulnerability management system when you automate these steps. You won't need to remember to run scans when you set up the system to automatically start them at regular intervals. Most likely, you'll scan during off hours when any additional load on the systems won't cause performance issues. The scans will then generate updated reports, which are emailed or placed in a shared network location for perusal at your convenience.

By scheduling scans to run on a regular basis and then automatically importing the results into your database, your vulnerability information will always be up-to-date. This process lets you safely automate reporting, because the weekly generated reports use fresh data. Similarly, by periodically updating your other data sources, such as Metasploit and the cve-search database, you can be confident that the third-party data you draw upon in your reports is also current.

In the scripts in Part II of this book, you'll leverage the standard Linux/Unix scheduling utility—the *cron* daemon—to automate the collection and the analysis of your vulnerability data. To coordinate all of the tasks, from data collection to report generation, you'll use shell scripts to run your Python scripts in sequence. By doing so, you'll prevent, for instance, the reporting script from running while the scanners are still collecting data about the environment. These scripts use a one-week interval, but your organization's collection and reporting interval will depend on how often you need a fresh view of your organization's vulnerability landscape.

Exploiting Your System's Vulnerabilities

At this point in your analysis, you have a regularly updated enterprise view that includes hosts, known vulnerabilities on those hosts, and any related known exploits that you could use against those hosts. From here, you can provide prioritized vulnerability information to system and application owners. You can also go one step further and attempt to exploit these vulnerabilities.

The first option is already a successful outcome of the vulnerability management process. The second option looks at the exploitable vulnerabilities list and runs a penetration test against affected hosts to determine whether they're exploitable. If successful, this option provides an additional level of prioritization to the results: not only is a system in principle exploitable, but it has been exploited.

There are two ways to attempt to exploit vulnerabilities. First, you can use a human penetration tester, either a security analyst with penetration testing skills or an outside auditor. Second, you can extend your automation by bringing Metasploit back into the process. Now, instead of just getting a list of exploits from it, you'll automate it to exploit those potentially exploitable hosts. This might seem like an excellent option or it might seem very frightening, depending on your perspective. Both perspectives are valid.

For those security analysts who have already seen the value of automating the vulnerability process, attempting to exploit your system might seem like a logical next step. You have a list of exploits and a list of vulnerable hosts, so why not check them out?

For more cautious analysts, exploiting their systems looks like a recipe for disaster. Running live exploits in a production environment is even more unpredictable than running scans: hosts could be taken down, networks could be clogged, and with only an automated system to blame, real heads might roll.

As with the rest of your security program, your decision depends on what you're trying to accomplish and your organization's risk tolerance. If your organization would rather incur a DoS attack than get attacked by way of an unpatched exploitable vulnerability, perhaps automated exploitation attempts are an option. On the other hand, if you're in a more risk-averse environment, tread very carefully: be sure to have full buy-in and acknowledgment of the risks from your CIO or the equivalent executive.

I'll briefly discuss how to integrate Metasploit into your vulnerability management program in this fashion in Chapter 14. But the actual process—particularly automation—will be left to you as an exercise. Automation is a powerful tool, but you must temper it with skill and extreme caution.

Summary

In this chapter, you learned how to take the raw vulnerability information from your scanners and shape it into usable intelligence. By combining data from your scanners with information about your network, additional information sources, and exploitability information, you can prioritize the vulnerabilities and focus on remediating the most severe issues.

In the next chapter, you'll learn how to remediate by patching and mitigating vulnerabilities as well as effecting systemic change to improve your organization's security posture.

5

DEALING WITH VULNERABILITIES

All the data collection and analysis you do will be for nothing if you don't have a clear goal in mind for how you'll use the results. In this chapter, you'll learn to use your vulnerability analysis to improve your organization's baseline security level. We'll look at three broad categories of security measures: patching, mitigation, and systemic measures. Patches and mitigations are direct responses and are almost always the most pressing. But the lasting value in any security program is systemic change brought about by improved intelligence. Although it might seem counterintuitive, another option is to accept the existing risk. I'll discuss why that could be the correct decision in your environment.

Security Measures

All of the measures mentioned you can apply in combination as part of a defense-in-depth strategy. Using this approach, you put redundant controls in place to defend against a vulnerability or category of vulnerabilities. Additionally, you must test these controls to confirm their effectiveness in protecting against vulnerability exploitation.

Patching

Patching is applying updates to resolve bugs and address vulnerabilities. Your first step after learning about new vulnerabilities is to determine the availability of a patch and apply it as soon as possible. You want to close the hole before any adversaries take advantage of it. Of course, this is easier said than done. Patching might be as easy as running an "updater" program on the system. Or it could be as complicated as compiling new code from an unofficial resource, applying it, and crossing your fingers. Patch management products, such as *Microsoft System Center Configuration Manager (SCCM),* are very helpful. But centralized patch tools might not be available for all operating systems and devices in your organization. In addition, there may be other reasons that patching is not feasible: perhaps the vulnerability is too new for the developer to have addressed it, the software is no longer being updated, or business requirements prioritize uptime over updates. Whatever the case, your next step is to look at how to mitigate the vulnerability without directly modifying the vulnerable application or OS.

Mitigation

Mitigations are a broad category of actions that either make exploitation more difficult or make the exploitation consequences less severe. For example, if you had a vulnerability on a specific Linux server daemon, you could use a firewall to block the port the daemon is using, preventing anyone from accessing it and exploiting the vulnerability. Of course, this isn't always a reasonable response. With the exception of network services that listen only on localhost (127.0.0.1) for local connections, most network services are running to communicate with outside systems. So if you're going to close them, you might as well turn them off. Instead, you might limit or block the vulnerable system from initiating connections to other internal systems. That way, even if the server is compromised, it will be difficult for the attacker to move laterally within the organization and compromise more systems.

Mitigations fall into a number of overlapping categories:

Application-based   These mitigations modify the vulnerable application to remove or limit the danger of exploitation. For example, if a vulnerability is in an Apache module that you're unable to patch immediately, you might modify the Apache configuration to disable that module. Alternatively, you could filter requests to the module and exclude known attack patterns.

Host-based This mitigation is performed at the OS level, rather than on the application. A host-based firewall or a system tool like SELinux, which limits the effectiveness of exploitation by making it harder to modify the underlying system, are good examples of host-based mitigations.

Network-based This mitigation is performed at the network level to intercept or monitor traffic to/from the vulnerable host. An example is a physical firewall, or an IDS, which watches for traffic indicating an attack or a successful compromise.

Temporary Some mitigations are more like bandages and are designed to be temporary. For example, blocking all access to a vulnerable host is probably not a lasting solution. But until you find a patch or a more permanent mitigation, it's invaluable.

Permanent If a mitigation doesn't affect the normal functionality of a vulnerable product, it often makes sense to leave it in place, even after you address the underlying vulnerability. This provides additional security against future vulnerabilities and improves the system's overall security posture.

Logical As distinguished from physical mitigations, logical mitigations occur at the software or network level (where most vulnerabilities exist).

Physical Sometimes you'll need physical mitigations. For instance, some secure environments prohibit removable USB drives to stop data exfiltration or the introduction of malware. Instead of disabling the USB ports in software, these organizations epoxy the ports shut, physically preventing the insertion of USB devices.

Systemic Measures

You implement a mitigation to protect against a specific vulnerability (or class of vulnerabilities), but you take systemic measures to improve the organization's overall security posture. Mitigations are reactive, whereas systemic measures are proactive. You might implement specific mitigations in response to a particular threat or vulnerability. But later, during a postmortem or a scheduled security review, the need for these mitigations might lead you to rethink your security posture to protect against future threats in that category.

Here's an example: you have a MySQL vulnerability on a system that you can't immediately patch. So you block access to the MySQL port (TCP 3306) using your internal firewall; modify the MySQL configuration to listen only on a local socket; and when you can, patch MySQL to a newer version. So far, you've done some mitigation (and a bit of patching). Now take a step back: why was that port open anyway? Maybe it should have been closed beforehand. What about other MySQL servers that weren't vulnerable to this specific issue? Do they really need to be listening for remote connections? What if that first system had been successfully compromised?

Could an attacker have used it as a launch point for further attacks? By thinking about these systemic questions, you can make broader changes that improve your overall security environment or at least start a conversation to influence organizational policies and configuration standards.

Accept the Risk

A fourth option is to do nothing at all. In risk management terms, this is called *risk acceptance*. Risk acceptance happens when you decide that the risk is so rare or so low impact that it's not worth addressing. There are cases where this is the best option. But even when you accept the risk, you need to document this lack of action and ensure that stakeholders agree with the decision. If auditors come around, you'll need to show them that you've considered and accepted the risk, not just neglected it. From the outside, an accepted but undocumented risk and an unnoticed risk look the same!

Defense in Depth

Most responses to a vulnerability will encompass patching, mitigation, and systemic change. *Defense in depth* is the practice of using multiple defensive measures in concert. Although you might not be able to apply a patch immediately on a critical system, you can quickly put some mitigations in place to block or at least limit the exploitation severity. Once you can patch the system, you can remove some of the more draconian mitigations (for instance, blocking all access to the affected service) while leaving some mitigations, such as improved configurations, in place. When certain mitigations are applied to multiple systems or across network environments, they might become systemic improvements.

This is the power of defense in depth: if you have multiple layers of protection, one layer can mitigate the failure of another layer. Even better, layered defenses can help protect against vulnerabilities that haven't been discovered. If a new MySQL zero-day vulnerability comes out that requires direct access, your existing measures of permitting access from only trusted hosts might be the difference between a day one exploitation and having enough time to apply the patch.

Validating Controls

Once you've put a patch, mitigation, or systemic change into place, the last step is to validate the presence and effectiveness of your chosen controls. In other words, you have to test those controls. With a mitigation or a systemic change, you can sometimes do this using additional scans: if the scanner no longer reports a vulnerability, the change is successful. Sometimes manual testing is best, especially if the mitigation or other change has subtle effects. Although a vulnerability scanner might have a hard time determining whether, for instance, you've disabled specific login options in Secure Shell (SSH), a human can craft test cases to validate this change.

Patching might seem like a simple case to validate. If you can confirm the patch has been applied, the vulnerability has been removed, right? But in reality, applying a patch doesn't always fully resolve the issue. Vendor-supplied patches (especially unofficial patches) might be incomplete, incorrectly applied, or missed on some systems, or they might cause other issues with the system and need to be rolled back. It's best to treat patching like any other mitigation and determine its effectiveness through active testing.

You can validate controls with the tools you already have by doing a vulnerability scan of a system before and after you apply controls. When you do, you should see a difference in the results. Let's say you have a remote code execution vulnerability on a MySQL deployment, and the system owner claims to have applied the patch to address this vulnerability. You would expect that a new scan of that system will show that the vulnerability no longer exists; if it does, you know the patch was incorrectly applied, not applied at all, or ineffective. If the mitigation was instead to restrict MySQL database connections to localhost, your scan should show that the port is no longer open on the system's public network interfaces.

But if your mitigation was to restrict MySQL connections to the local network, your scan will show nothing different. In such situations, devise testing methods based on the purpose of the mitigations or systemic measures that have been put in place. In this example, you might run another scan from outside your local network segment.

Although a full discussion of control validation techniques is outside the scope of the present topic, the general principle is simple: don't assume anything works! Always test to make sure that any measures you've taken to address the vulnerability are addressing it.

Summary

In this chapter, we took a high-level look at the actions you can take when you discover vulnerabilities in your environment. Whether immediate, systemic, or—in the case of risk acceptance—just in the form of documentation, the measures you take to protect against vulnerabilities will depend on the system you're working within.

In the next chapter, you'll learn how to work within your organization's structures to put these measures into effect and bring about lasting improvements to your system's security posture.

6

ORGANIZATIONAL SUPPORT AND OFFICE POLITICS

Once you set up your vulnerability management environment and scripts, you can obtain valuable information about exploitable vulnerabilities on devices across your network. You can also act on the information in a way that will improve your organization's security posture. If you work in a small organization, you're probably the security analyst *and* the systems administrator. So it's up to you to patch your vulnerable systems, and you're done. But if you work in a larger environment, the problem is more complicated and business oriented. You'll have to interact with other IT staff. In addition, you'll likely need to know the next update window, who performs the updates, and, most pressingly, what motivation the system or application owners have to apply the patches that will resolve the vulnerabilities you've found.

This chapter focuses on the human-data interface. To use your analysis results to effect real security improvement in your organization's environment, you need a solid grasp of interpersonal interactions and your organization's structure and politics. No scripts exist for this more humancentric portion of the vulnerability management process. But there are some guidelines to help make this process as smooth as possible.

Balancing Competing Priorities

Being a security analyst can be difficult. You see what needs to be done to improve an organization's security, but you can't always resolve the issues yourself. Instead, you must work with other functional groups and individuals who might not share your priorities. Everyone does their part to keep the business running smoothly: you improve security, systems administrators and database administrators (DBAs) increase uptime and keep the systems running, and application owners ensure that their programs continue to function properly. On the macro level, everyone's goals align. But in the trenches, these differing goals can conflict.

Imagine a security analyst who's in possession of a short list of vulnerabilities in the organization that are, according to their standards, critical. Among them is a known vulnerability in the Windows kernel that attackers can exploit remotely and without authentication. This vulnerability exists in a database server that is vital to the organization's operations. A Windows administrator is in charge of the box, a DBA manages the SQL Server instance on the box, and a manager is in charge of the application using that database.

In this situation, the analyst should consider three factors. First is the organization's structure. If all three people they need to talk to share a common manager, the analyst should initially talk to the manager. Second, before doing so, they should consider the organization's politics. It might be impolitic to speak to an executive directly. If so, the next step is to escalate the issue up their own chain of command. Third, and most important, the analyst should consider the official policy. If there is a stated policy for security issues, they're in luck: they follow the policy. If the organization lacks a policy, they should consider getting a policy. It won't help them now when they have a critical vulnerability on a business-critical database server, but it'll set an important precedent for the future.

The security analyst wants to address the vulnerability to remove an easy attack point and improve the organization's overall security posture. But think about this issue from the systems administrator's point of view. Applying a patch to the Windows kernel means that the system will probably have to be rebooted. A reboot means downtime. Also, if any issues occur with the patch—and Microsoft has more than once produced patches that cause serious issues or even a full system breakdown when applied—more downtime could happen and maybe require a hands-on recovery (even more downtime!). Even if the recommendation is for mitigation strategies, such as locking down services or blocking ports, these measures might lead to unforeseen system problems.

The system administrator is very cautious about agreeing to address the issue, because it goes against their immediate priorities. Likewise, the DBA and application owner will be concerned about the downtime and any unforeseen consequences of the patch or mitigation. If all three are united against taking any action, the issue is likely to persist for some time, if not indefinitely. Again, none of these priorities are wrong. But the ground-level tactics used to achieve individual priorities are in conflict.

Gaining Support

In the long term, the answer to everyone's issues is straightforward: strong information security and risk management governance. With the proper policies and authority structures in place, the organization can align its priorities and address issues in a way that is consistent with proper risk management strategies. If the priority is uptime, patching might have to wait. If the priority is system security, uptime will take a back seat.

But that's a discussion for another time (and another pay grade). In many organizations, these processes and policies are incomplete or non-existent. You'll still have to find a way to resolve problems. In the following sections, I provide strategies that can help you, as a security analyst, navigate any organizational barriers to address the priorities that you've identified in your vulnerability management processes.

Empathy

We can get so caught up in our own ideas of how to protect the organization that we forget that other people have their own, equally valid perspectives. When trying to convince others of your point of view, put yourself in their shoes and try to understand what they want. When talking to a Windows administrator who must keep their systems up and running, recognize this and remind them that although patching incurs some downtime, you can control that downtime and do it at an appropriate time. Likewise, remind them that if the server is compromised through an unpatched vulnerability, the downtime will be unexpected and potentially protracted. The DBA and application owner are concerned about data integrity and confidentiality. Emphasizing that a vulnerability might lead to data manipulation, destruction, or disclosure might be enough to convince them that improved data protection is worth some downtime.

Empathizing with others won't always do the trick. Sometimes you'll need to take additional steps. But by starting with an ounce of understanding and willingness to listen to someone else's viewpoint, you might save yourself a world of pain. You might even gain a new ally in the cause of increased security.

Involve Stakeholders Early

We all know how it feels to have decisions forced on us without being involved in the decision-making process, particularly when rulings might impact us negatively. Scans and vulnerability remediation affect system administrators, application developers, application owners, and end users, among others. Instead of conducting the entire vulnerability management process on your own and then emerging with a list of directives, involve others early in the process. If you're responsive to stakeholders' concerns, they'll likely be more supportive of your recommendations or directives, even if from time to time you must overrule their desires.

Understand Office Politics

Often, technical people try to skirt office politics. They'd rather get their work done without worrying about who is speaking to whom, who just won a big contract, who has pull with the executives, and who doesn't. But ignoring these issues doesn't make them go away: they're real and powerful. Knowing who you should approach and how you should approach them will help you get approval for difficult or controversial actions (like taking a server offline to patch). It's also important to know whose toes you're stepping on when you push such actions. Sure, you might win the immediate battle of applying that patch or mitigation, but in a few months, you might find that you've made an enemy of a systems manager who's now stonewalling you at every turn.

In many ways, this office politics strategy is simply empathy taken to another level. In some respects, even technical work is social. Often, the changes you'll want to implement have a social impact. Understanding an organization's official and unofficial structures, and then imagining yourself as a co-worker navigating those structures, can help you appreciate your colleagues' perspectives on your proposed actions.

Speak Their Language

Empathizing with someone goes hand in hand with being able to speak to them in terms they understand. A technical discussion about uptime, patch levels, and exploits makes sense to technical staff, such as server administrators and DBAs. But if that same discussion happens with more business-oriented staff, they might tune out as soon as you start talking. Learn about the kinds of concerns they have and the metrics they use, then couch your conversations in those terms if possible. An application owner or others in a business-oriented role might be more concerned about return on investment or, if they're involved in risk management, risk/control terminology. By phrasing your arguments using their preferred terms, you can show an awareness of their concerns and your willingness to recognize other perspectives when coming to your conclusions.

Find a Champion

Having the approval of a C-level executive interested in the organization's security might be the most effective way to get support for security-related activities. But if you're fortunate enough to have the ear of a CIO, CTO, or even a CFO or audit officer, don't overuse that privilege: security concerns aren't their only priority. In addition, be aware that if you're constantly going over your co-workers' heads to get your way, it will only cause more friction. On the other hand, if your organization has a chief information security officer (CISO), you're in a strong position to ensure that security concerns are heard and understood at a high organizational level.

Argue for Risk Management

You might not have a full risk management program in place. But that shouldn't stop you from taking a page (or several pages) from the risk management playbook when you're making your arguments to address vulnerabilities. You can use this simple formula in almost any situation:

$$\text{Risk} = \text{Likelihood} \times \text{Cost}$$

To estimate the overall risk of an adverse event, you need to look at two dimensions. First, how likely is it to occur? Second, if it does occur, how costly will that event be to your organization? Let's use the previously mentioned example of the known vulnerability in your database server. Let's also say you're working for a company with data that attackers are likely to want. The cost of an attempted breach happening is extremely high, and the probability of its happening is also high, especially if that proprietary data is on a vulnerable server. So the overall risk if that vulnerability is not addressed is very high. Conversely, if you do patch the vulnerability, you need to look at the likelihood of a bad outcome (crash, application incompatibility, and so on) and the cost of those events (cleanup time, employee time addressing the issue, cost of potential lost business) to estimate the risk of fixing the problem. Once you have those two estimates, you can compare the relative risks of patching and not patching and decide what makes sense to all parties.

A simple way of calculating risk without using specific numbers is to use a *risk matrix*, like the one in Figure 6-1. A risk matrix is a simple table that ranks risks by assigning them relative likelihoods and costs, making it easier to understand the relative values of two otherwise dissimilar risks. Instead of trying to figure out exact percentages of likelihood and precise costs, you rank both on a scale of 1 to 5. Then you combine the two axes to determine the overall risk level—low to high. You can involve other stakeholders in these risk calculations to ensure a consensus on the relative likelihoods and costs of various courses of action (this will also make explaining the risks to them much easier).

Likelihood/cost	1	2	3	4	5
1	1	2	3	4	5
2	2	4	6	8	10
3	3	6	9	12	15
4	4	8	12	16	20
5	5	10	15	20	25

☐ Low risk (1–7)   ■ Medium risk (8–14)   ■ High risk (15+)

Figure 6-1: Simple risk calculation

Continuing with the example, let's look at the two courses of action—patch or don't patch—in these risk matrix terms. If we patch, what is the likelihood of issues? Well, historically in this environment, there is a small chance of patches causing issues on Windows systems, so we'll rank it as a 2 (to be conservative). If issues happen, what will be the cost of that event? Well, it will consume administrator time, possibly requiring a patch rollback or a restore from backup. At the same time, we might lose several hours of processing because this is a critical database. On a scale of 1 (negligible) to 5 (catastrophic), this patching problem would be painful but relatively brief. Let's give it a risk of 3. So the overall risk is 6, putting it in the "Low risk" category.

Now let's look at not patching. What are the odds of someone compromising the system if we don't patch? It's a known vulnerability with known exploits, so that makes it more likely to be compromised than a vulnerability with no known exploits. But it's in a protected network segment, a circumstance that provides some mitigation. We'll rank it as 3 for likelihood. What about the cost? The data on that database is confidential, and the application it runs is critical to company operations. If the database is simply brought down, we can restore it from a backup, as in the other scenario. But if that data is compromised, it could be very serious: we would lose competitive advantage and could face significant negative publicity if it becomes public knowledge that we were attacked. We'll give it a cost of 5. The overall risk is now 15. The risk of taking no action is much higher than the risk of applying the patch, so taking action is the right thing to do.

You can use these risk calculations for more than simple act/don't act decisions. Although a complete discussion of information security risk management is not the goal, here are two other situations in which the risk calculations might be very useful:

- **Weighing the cost/benefit of a new piece of security hardware or software:** If a tool costs the company $150,000 but not addressing the risks it mitigates is likely to cost only $10,000, it's not a good investment.

- **Estimating the damage that could be done by various types of adversaries and considering available countermeasures:** If the attackers are opportunistic criminals, it might make sense to focus defenses against the less sophisticated attackers first (or exclusively) to maximize security dollars. If the attackers are advanced black hat hackers, more sophisticated (and expensive) countermeasures will be necessary.

Summary

Security practitioners often have great responsibility coupled with little to no formal authority. Working against bureaucracy can sometimes feel like shoveling back the tide. But by building relationships within the organization and making a strong case for enacting security measures (by patching, mitigation, or more systemic changes), you can accomplish more than you would by treating every new security incident as an isolated battle to fight against the rest of your organization. Gaining executive sponsorship for an IT risk management program will pay dividends in the long run, and you can leverage risk management concepts to make an even stronger case.

This is the end of Part I. You've taken a whistle-stop tour of the vulnerability management practice and its tasks and components. You also learned how to take effective steps to improve your organization's vulnerability posture. In Part II, you'll build the vulnerability management system we've been discussing.

PART II

HANDS-ON VULNERABILITY MANAGEMENT

7

SETTING UP YOUR ENVIRONMENT

To start constructing a complete vulnerability management system, you must first
build the foundation. In this chapter, you'll
set up the base Linux environment for your
system, install the tools you'll use in the following
chapters, and write a script to ensure all the components are regularly updated.

Setting Up the System

Your first step, of course, is to set up a Linux-based environment for the
basic tools you'll use to build your system: those tools include Nmap,
OpenVAS, cve-search, and Metasploit.

> **HARDWARE PREREQUISITES**
>
> For a small network, you'll only need one CPU and 4GB of RAM. But for larger environments, you should add more CPUs and 8GB or more of RAM. Storage-wise, 50GB of disk space will suffice in a small environment, but large MongoDB instances will quickly fill disk space. So, consider upgrading to 250GB of disk space or more.

Installing the OS and Packages

We'll use the following software in this guide. Many of these packages will be pre-installed in most Linux distributions. Install the rest using the package manager that comes with your system. (Ubuntu and other Debian-based systems use *apt*, whereas Redhat-based distributions typically use *yum*.)

- Linux (this guide uses Ubuntu 18.04 LTS)
- Python 3.3 or later (cve-search requires this Python version)
- MongoDB 2.2 or later, and MongoDB development headers (the package names will vary by distribution; Ubuntu uses mongodb and mongodb-dev)
- SQLite 3 (OpenVAS requires this version)
- Nmap
- pip3 (to install additional Python packages; package is python3-pip in Ubuntu)
- Git (for cve-search, Metasploit, Exploit-db)
- libxml2-dev, libxslt1-dev, zlib1g-dev (for cve-search)
- jq (for JSON parsing)
- cURL (to download files and scripts)
- psql (the PostgreSQL client; to manually look at the Metasploit Framework database)

The Linux distribution you use is up to you. Using a prebuilt virtual machine image can save some time, although I recommend working through the Linux install process manually to customize the installation to your needs.

Although it's not a requirement, I recommend setting up a dedicated user (I used vmadmin) with sudo privileges that you'll use to run the scripts we will create. You'll increase your system security by not using the root user.

Customize It

To tailor this basic setup to your own environment, you could add more scanners (on separate physical or virtual hardware) and configure them to send their reports to a centralized location. Or you could locate the MongoDB instance on a separate machine or a shared server. Using databases other than MongoDB is also an option, although doing so will require modifying the scripts you write.

By deploying these packages on a compact single-board system, such as the Raspberry Pi, you can make a palm-sized vulnerability-scanning system that you could use at different locations as a portable vulnerability management tool.

You can also use a different Unix-type operating system, such as BSD or a commercial Unix like Solaris, or even Windows with the appropriate tools (Python, MongoDB) in place. But collecting all the prerequisites might be a chore. Recent versions of Windows 10 with the Linux subsystem installed can use standard Ubuntu packages.

Installing the Tools

After setting up your base Linux system, the next step is to install the main tools—OpenVAS, cve-search, and Metasploit—which you'll use to build your vulnerability management system.

 In this book, the # prompt indicates that you must run that command as root or with root privileges via sudo.

Setting Up OpenVAS

OpenVAS is an open source vulnerability scanner that derived from Nessus when Nessus became closed source. The OpenVAS community and Greenbone Networks GmbH currently maintain it.

Installing the Packages

OpenVAS is not part of the standard Ubuntu repository, so we need to add a custom repo to the trusted Ubuntu list. We'll use the OpenVAS repository built by Mohammad Razavi, which is available at *https://launchpad.net/~mrazavi/+archive/ubuntu/openvas/*.

To get these packages, add this repository information to your *apt* software sources:

```
# add-apt-repository ppa:mrazavi/openvas
```

Next, update *apt* to make *apt* aware that the OpenVAS software is in this custom repository:

```
# apt-get update
```

Download and install OpenVAS. The download consists of about 100MB of data and in total will take up approximately 500MB of space:

```
# apt-get install openvas9
```

Updating OpenVAS

Once the package is installed, run the setup scripts in Listing 7-1 to sync data that OpenVAS uses in its scans. You must run all these scripts as root.

```
  # greenbone-nvt-sync
  # greenbone-scapdata-sync
  # greenbone-certdata-sync
  # service openvas-scanner restart
  # service openvas-manager restart
❶ # openvasmd --rebuild --progress
```

Listing 7-1: Getting OpenVAS ready to go

The rebuilding tool ❶ might take a long time to complete. This is normal and is nothing to worry about.

Edit your Redis configuration, which the OpenVAS scanner daemon uses for temporary result storage. Edit /etc/redis/redis.conf and comment out any lines with the format **save xx yy** (for example, **save 900 1**). Then restart Redis (**# service redis-server restart**) and the OpenVAS scanner (**# service openvas-scanner restart**).

NOTE   *The preceding instructions are based on OpenVAS version 9 on Ubuntu. Future versions might have different installation and update instructions.*

Test the Deployment

Go to *https://<your-ip-address>:4000*. If everything is up and running, you should see a Greenbone Security Assistant login page. The default login to Greenbone is **admin**/**admin**. Once you're logged in, click around to familiarize yourself with the interface—and don't forget to change the default credentials to something more secure. Setting up and running scans is a bit complicated, but you'll delve into more of the OpenVAS and Greenbone scan options in Chapter 8.

Setting Up cve-search

The tool cve-search is a set of Python scripts backed by a MongoDB database that contains a vast amount of publicly available vulnerability information from the official CVE database at *https://cve.mitre.org/*. You'll mostly use the cve-search database, not the frontend tools, but those utilities might be useful for doing manual vulnerability searches.

Downloading cve-search

You can download cve-search as a *.zip* or *.tarball* file from the cve-search site (*https://adulau.github.io/cve-search/*) and extract it or get it directly from the developer's online repository using Git. Unlike many of the commands in these chapters, you can do this as a normal (unprivileged) user. The following command installs the cve-search tool into the *./cve-search* directory:

```
$ git clone https://github.com/cve-search/cve-search.git
```

Installing Dependencies via pip

To ensure that all of cve-search's prerequisites are in place, use the pip3 tool to set up the requirements in *requirements.txt* (a file that comes with cve-search). Before running this command, ensure that you've installed the libxml2-dev, libxslt1-dev, and zlib1g-dev packages or the equivalent for your Linux distribution:

```
$ cd cve-search; sudo pip3 install -r requirements.txt
```

If this command fails, don't despair. Be patient, look at where errors are occurring in the installation process, make any necessary changes, and retry. You might need to add additional packages via your distribution's package manager.

Populating the Database

Finally, you'll build and populate the MongoDB database that serves as the datastore for the cve-search tools using the commands in Listing 7-2. The second and third scripts might take quite a long time to complete.

```
$ ./sbin/db_mgmt_json.py -p
Database population started
Importing CVEs for year 2002
Importing CVEs for year 2003
Importing CVEs for year 2004
Importing CVEs for year 2005
--snip--
$ ./sbin/db_mgmt_cpe_dictionary.py
Preparing [################################################] 194571/194571
$ ./sbin/db_updater.py -c
INFO:root:Starting cve
Preparing [################################################] 630/630
INFO:root:cve has 120714 elements (0 update)
INFO:root:Starting cpe
Not modified
--snip--
INFO:root:
[-] No plugin loader file!
```

Listing 7-2: Building and updating the CVE database

Testing cve-search

Once the tools are installed and the database is populated, try a simple search to see what kind of information is available in the CVE database. The command `./bin/search.py -c CVE-2010-3333 -o json|jq` will find information about CVE-2010-3333, a stack buffer overflow vulnerability in Microsoft Office. Piping the command through jq will format the JSON blob into something much more readable. We'll look at CVE information in more detail in Chapter 8.

Setting Up Metasploit

The Metasploit Framework tool contains numerous working exploits as well as a scriptable Ruby environment for automating repetitive or complex exploitation tasks. This step is optional because you can build nearly the entire system without using Metasploit.

Installing the Metasploit Framework

The easiest way to deploy the Metasploit Framework on a Linux system is to use the installer script detailed at *https://github.com/rapid7/metasploit -framework/wiki/Nightly-Installers/*. This script sets up the proper Metasploit repositories and integrates with package management systems, such as *yum* and *apt*. Another option is to clone the Git repository and build directly from there, but I follow the installer method in this deployment.

Execute this long chain of commands as root to download and run the latest version of the installer script (*msfinstall*), which will add the Metasploit repository to *apt* and then install the Metasploit Framework:

```
# curl https://raw.githubusercontent.com/rapid7/metasploit-omnibus/master/config/templates
/metasploit-framework-wrappers/msfupdate.erb > msfinstall && chmod 755 msfinstall &&
./msfinstall
```

If you want more control over the individual steps, follow the instructions at the Github URL above.

Completing and Testing the Installation

Run `msfconsole` as root to complete the setup.

```
# msfconsole
```

This step sets up the Postgres database and eventually brings you to an `msf>` prompt. From there, you can explore the Metasploit Framework. For example, if you're interested in whether Metasploit has exploits for the 2014 "Heartbleed" vulnerability, search for `CVE-2014-0160`, as shown in Listing 7-3.

```
msf > search cve-2014-0160
[!] Module database cache not built yet, using slow search
```

```
Matching Modules
================

  #  Name  Disclosure Date  Rank  Check  Description
  -  ----  ---------------  ----  -----  -----------
❶ 0  auxiliary/scanner/ssl/openssl_heartbleed        2014-04-07
   normal  Yes    OpenSSL Heartbeat (Heartbleed) Information Leak
❷ 1  auxiliary/server/openssl_heartbeat_client_memory 2014-04-07
   normal  No     OpenSSL Heartbeat (Heartbleed) Client Memory Exposure
```

Listing 7-3: Sample Metasploit search output

Either of the two exploits ❶❷ matching this vulnerability could exploit
a system vulnerable to CVE-2014-0160. Because I'm just covering installing the Metasploit Framework, I'll leave further exploitation for you as an
exercise. Keep in mind that Metasploit includes working exploit code and
it can cause serious issues, including but not limited to crashes, on any systems you try attacking with Metasploit. Be careful and try it only on systems
you're authorized to attack.

Customize It

If you prefer or if your Linux distribution doesn't have prebuilt packages
for OpenVAS available, you can install OpenVAS from source. Doing so will
give you some opportunities to customize it for your environment.

Although I installed OpenVAS 9 in this chapter, earlier versions are
functional (and might be easier to find for your architecture). Those versions might have a different XML output format, which means you'll need
to modify your scripts.

You can install the cve-search set of scripts in a different location (by
cloning the Git repository into a different location) if you want multiple
users to use the tool.

By installing Metasploit manually, you'll have more control over the
additional packages it uses—Ruby and PostgreSQL in particular—and can
customize your deployment more fully.

Keeping the System Updated

Once all the software is installed, you'll need to keep it up-to-date. The
script in this section shows you how to create an updater script that is suit-
able for running on a schedule.

Writing a Script for Automatic Updates

First, create a bash script that will run a number of update scripts for the
tools you've deployed. This way, instead of having to run numerous separate
update scripts every time you want to update your toolchain, you can run
a single script. Even better, you can add this script to the system scheduler
(*cron*) so it will run automatically on a regular basis.

Save Listing 7-4 as *update-vm-tools.sh*.

```
❶ #!/bin/bash
❷ CVE_SEARCH_DIR=/path/to/cve-search

❸ LOG=/path/to/output.log

  # this clears the log file by overwriting it with a single
  # line containing the date and time to an empty file
❹ date > ${LOG}

❺ greenbone-nvt-sync >> ${LOG}
  greenbone-scapdata-sync >> ${LOG}
  greenbone-certdata-sync >> ${LOG}
  service openvas-scanner restart >> ${LOG}
  service openvas-manager restart >> ${LOG}
  openvasmd --rebuild >> ${LOG}

❻ ${CVE_SEARCH_DIR}/sbin/db_updater.py -v >> ${LOG}

❼ apt-get -y update >> ${LOG}

❽ msfupdate >> ${LOG}

❾ echo Update process done. >> ${LOG}
```

Listing 7-4: A simple system updater script

The shebang identifies that this is a bash shell script ❶, so when you run it, the system knows which interpreter to use. The CVE_SEARCH_DIR variable ❷ points to the cve-search path on your system. The LOG variable ❸ points to a log file, which starts with the current date ❹. The output of all the update commands will be written to the log file.

The same commands you used in Listing 7-1 to sync OpenVAS are used to update it ❺. Check the specific version of OpenVAS that you've installed to ensure that these executables and paths are correct. Then execute the cve-search updater using the value stored in CVE_SEARCH_DIR to refer to the actual cve-search path on your system ❻. To update the underlying Linux system as well as the OpenVAS packages, run a full system update using the -y flag so the update doesn't ask for confirmation and runs without human interaction ❼. (If you're running a Linux system that isn't based on Debian, your update command might be different; for instance, RPM-based systems like Redhat use *yum* to update.) Next, we update Metasploit using its own msfupdate script ❽ and write Update process done to the log file ❾.

Set the script as an executable using the following chmod command:

```
# chmod +x update-vm-tools.sh
```

Now you can run the script at any time to update your vulnerability management toolchain.

Running the Script Automatically

Now that you have a single update script, you can add it to *crontab* to run regularly. I set it up to run weekly on Sundays at 4 AM, but you can change that to a different time.

Edit */etc/crontab* (as root) and add the following line at the bottom of the file:

```
0 4 * * 7 root /path/to/update-vm-tools.sh
```

Many Linux distributions have directories that will automatically run anything in them on a regular basis using *cron*. For example, Ubuntu runs any script placed in the */etc/cron.weekly/* directory at 6 AM on Sundays. If you want to use this method instead, simply save your update script in that directory or create a symlink to it.

To ensure that the script ran successfully, you can look at the log file generated in the update script (*/path/to/update.log* in Listing 7-4 ❸) and look at the output of each update script.

Customize It

I placed the data update and system update components in a single script for simplicity. But you might prefer to separate data and application updates either by making the *apt* update manual or by running it on a different schedule from the other update scripts. You might also change the `apt-get` command to update only Metasploit and OpenVAS, saving the full system updates for a manual process.

In later chapters (particularly Chapter 12), you'll schedule other scripts for data collection and analysis. When setting the specific timeframes for these other scripts to run, keep in mind that they might conflict with the system updates in this script, so schedule accordingly.

Summary

In this chapter, you took the first steps to building a complete vulnerability management system: you set up the OS and the underlying tools that you'll be writing script controls for in the following chapters. It might not seem like much, but from small beginnings you'll build great things.

In the next chapter, you'll look at Nmap, cve-search, and OpenVAS in more detail so you can become familiar with their features before you start controlling them via shell scripts and Python.

8

USING THE DATA COLLECTION TOOLS

The goal for your vulnerability management system is to have usable vulnerability data in a database to make it easy to search, analyze, and generate automated reports. You've installed all the basic tools, but your database is empty.

In this chapter, you'll look at the tools you'll use to collect the raw data for your vulnerability management system: they include Nmap, OpenVAS, and cve-search. To gain some basic familiarity with each tool, you'll run each manually to explore its configuration options and see what kind of data you can collect with it. If you're already familiar with these tools, you can move on to the next chapter. There you'll collect the information you need and store it in the database.

An Introduction to the Tools

Although the information you can gather with the Nmap, OpenVAS, and cve-search tools overlaps, the purpose of each tool is very different. Before we get into the nitty-gritty of command line options and XML outputs, let's situate the three of them in the overall program.

Nmap

The Nmap network discovery scanning tool was originally developed in 1997 by Gordon Lyon, a programmer known in the security community as Fyodor. In its more than 20-year life, Nmap has had regular development and improvement, and it remains a central tool in the security specialist's tool belt.

How Does It Work?

The Nmap tool sends various packets to an IP address or a range of IP addresses to gather networking-related information about the hosts at those addresses. You can configure the specific packets it sends and which ports it sends them to: Nmap can be slow and stealthy or very fast and aggressive, depending on how you configure it.

What Is It Good For?

You can use Nmap to do quick scans of network ranges to discover information about the live hosts: their addresses, which services they're providing on which network ports, and which OS they're running.

What Is It Not Good For?

Although Nmap will tell you which ports are open on a host, it isn't a vulnerability scanner. In some cases, Nmap can determine the specific version of a server running on a port, but it won't match that information with known vulnerabilities or perform additional testing to determine whether certain vulnerabilities exist. The Nmap tool isn't sufficient for determining how at-risk any of the hosts it finds might be. Instead, it's best to use the information gleaned from Nmap and as the basis for further analysis, either on its own or with additional data from vulnerability scanners.

OpenVAS

OpenVAS is a vulnerability scanner. Its roots lie in an earlier scanner, Nessus, first released in 1998. In 2005, Tenable Network Security, founded by the Nessus core developers, converted Nessus into a commercial product. In response, a number of open source developers forked the Nessus codebase to continue providing a free and open source scanner project. The current result is OpenVAS, which only barely resembles the modern incarnation of the commercial Nessus product. OpenVAS is divided into a number of separate components: the scanner, a management and scheduling daemon, and the Greenbone Security Assistant, which is a web-based frontend to make configuring and running scans easy.

How Does It Work?

Like Nmap, OpenVAS sends a series of network packets to one or more IP addresses. But unlike Nmap, OpenVAS sends targeted packets to determine specific versions of services and whether or not they're vulnerable to known attacks. It also contains numerous plug-ins that do more aggressive testing for specific vulnerabilities beyond simple version checking. (For more details about the capabilities of vulnerability scanners, see Chapter 3.)

What Is It Good For?

OpenVAS is ideal for finding specific vulnerabilities on hosts in the targeted network. But, like any network vulnerability scanner, it will only discover network vulnerabilities, not those that can only be exploited from the local system. Its internal database includes significant background detail about the vulnerabilities it can find, including CVSS scores, exploitation consequences, and a comprehensive list of external references containing more vulnerability details. You can also use OpenVAS for host discovery.

What Is It Less Good For?

Although OpenVAS provides host discovery, its OS fingerprinting functionality is less complete than Nmap's, so it might not identify the specific OS the target is running. By running the two in concert, you'll get a more accurate view of your network environment.

Differences Between OpenVAS and Commercial Scanners

To be frank, the OpenVAS scanner is free and it shows. Its interfaces, both command line and web based, are functional and complete, but their usability leaves a lot to be desired. Without a commercially supported budget to develop new vulnerability tests, development of OpenVAS is largely left to the open source community. Older vulnerabilities are well supported, but new vulnerability coverage is significantly less comprehensive. However, in an environment where there is no budget for a vulnerability management program, this tool provides significant value. That said, this is the first component you should replace if you have a budget of $1,000 or more and can purchase a commercial scanning product, such as Nessus or Qualys.

cve-search

Once you have asset and vulnerability information from Nmap and OpenVAS, the next step is to bring in additional data sources. The cve-search tool suite will give you a comprehensive local repository of CVE data.

CVE/NVD

The CVE list, provided by the Mitre Corporation, is the authoritative source of CVE information. Its content is synchronized to the National Vulnerability Database (NVD) which is run by NIST; NVD adds additional vulnerability information, such as vulnerability rating, patch information, and more searching options. If you're looking for information on vulnerabilities, check these two lists first. They provide online search functionality, vulnerability feeds, and the option to mirror their databases locally. But if you want to programmatically grab and parse CVE information, the cve-search tools will save you a lot of time.

cve-search

The cve-search suite is a collection of Python scripts that gathers vulnerability information from several online sources (primarily the CVE/NVD repositories). It then puts that data into a Mongo database for easy querying and analysis. This suite is a gold mine for those interested in doing automated vulnerability data analysis. We'll use the cve-search suite to build and maintain our local CVE database. Then we'll query that database directly for analysis instead of using the cve-search frontend tools. But keep in mind that those tools are useful for manual vulnerability searching.

Getting Started with Nmap Scanning

The goal of this exercise is to gain a basic familiarity with Nmap scans. By using this tool, you can determine:

- The number of hosts in a given network segment
- The MAC address of each host, which you can leverage to figure out the underlying hardware
- Which ports are open on each host and which services are running on those ports
- Which OS might be running on the host

CAUTION

For all scripts that involve live scanning, you need a network range or series of hosts that you have permission to scan. Scanning with Nmap, although generally not intrusive, can cause problems with the scanned systems. The risks range from temporary slowness due to resource exhaustion to more serious issues that might require a system reboot or other manual intervention. Penetration testers regularly run into issues scanning networked printers: the probes may cause the printers to output page after page of garbage output, wasting reams of paper if the testers haven't excluded the printers or removed the paper in advance. Only scan systems you own or have permission to scan.

Although you'll look at a few scanning and output options, you'll only scratch the surface of Nmap's capabilities. I highly recommend reading the Nmap manual (run man nmap at the command prompt) to learn about all its features.

Running a Basic Scan

The best way to understand how Nmap works and what kind of information you can gather is to do a basic scan of your network range. I used the 10.0.1.0/24 range, which is my local test network. Running Nmap without any arguments except for the scan range makes it do a basic scan:

```
# nmap 10.0.1.0/24
```

This command scans the 1,000 most common ports on each address in the target network range. The output will look something like Listing 8-1.

NOTE *Although an unprivileged user can run this scan, you can gather more information when you run it as root. For example, the MAC address output in Listing 8-1 won't be available if you run an unprivileged scan.*

```
--snip--
❶ Nmap scan report for 10.0.1.4
❷ Host is up (0.0051s latency).
  Not shown: 997 filtered ports
❸ PORT     STATE SERVICE
  22/tcp   open  ssh
  88/tcp   open  kerberos-sec
  5900/tcp open  vnc
❹ MAC Address: B8:E8:56:15:68:20 (Apple)

  Nmap scan report for 10.0.1.5
  Host is up (0.0032s latency).
  Not shown: 996 filtered ports
  PORT     STATE SERVICE
  135/tcp  open  msrpc
  139/tcp  open  netbios-ssn
  445/tcp  open  microsoft-ds
  5357/tcp open  wsdapi
  MAC Address: 70:85:C2:4A:A9:90 (ASRock Incorporation)

  --snip--
❺ Nmap done: 256 IP addresses (7 hosts up) scanned in 663.38 seconds
  --snip--
```

Listing 8-1: Default nmap output when run as the root user

The output is separated by host ❶. For each host, you'll see whether the host is up and how long it takes to contact it ❷, the open ports and the service running on each port ❸, and the MAC (network hardware) address ❹ if available. If you run Nmap as an unprivileged user or if the targeted device is on another subnet, the MAC information won't be available. Additionally, there's a summary of scanned hosts and statistics for how long it took Nmap to complete its scan ❺.

Even just using the default options, you can learn a good deal about the systems you scan. However, Nmap can provide even more data.

Using Nmap Flags

You can use the nmap command's flags to learn more about your network or adjust your scans' output format.

Getting More Information with -v

The -v flag adds verbosity to the scan; this means that Nmap becomes chattier about what it's doing:

```
# nmap -v 10.0.1.0/24
```

You can add up to three v flags in a row (or -v3) to get more information that you can use to monitor the scan's progress. For example, adding -vvv to the same scan will result in Listing 8-2.

```
# nmap -vvv 10.0.1.0/24
Starting Nmap 7.01 ( https://nmap.org ) at 2020-03-04 10:42 PST
Initiating ARP Ping Scan at 10:42
Scanning 255 hosts [1 port/host]
adjust_timeouts2: packet supposedly had rtt of -137778 microseconds.  Ignoring time.
adjust_timeouts2: packet supposedly had rtt of -132660 microseconds.  Ignoring time.
adjust_timeouts2: packet supposedly had rtt of -54309 microseconds.  Ignoring time.
adjust_timeouts2: packet supposedly had rtt of -59003 microseconds.  Ignoring time.
adjust_timeouts2: packet supposedly had rtt of -59050 microseconds.  Ignoring time.
Completed ARP Ping Scan at 10:43, 3.26s elapsed (255 total hosts)
Initiating Parallel DNS resolution of 255 hosts. at 10:43
Completed Parallel DNS resolution of 255 hosts. at 10:43, 0.02s elapsed
DNS resolution of 16 IPs took 0.02s. Mode: Async [#: 1, OK: 11, NX: 5, DR: 0, SF: 0, TR: 16,
CN: 0]
Nmap scan report for 10.0.1.0 [host down, received no-response]
Nmap scan report for 10.0.1.2 [host down, received no-response]
Nmap scan report for 10.0.1.3 [host down, received no-response]
Nmap scan report for 10.0.1.4 [host down, received no-response]
--snip--
SYN Stealth Scan Timing: About 10.98% done; ETC: 13:37 (0:04:11 remaining)
Increasing send delay for 10.0.1.7 from 40 to 80 due to 11 out of 27 dropped probes since last
increase.
Increasing send delay for 10.0.1.18 from 40 to 80 due to 11 out of 23 dropped probes since last
increase.
--snip--
```

Listing 8-2: Verbose nmap output showing debugging and progress-related information

Under normal circumstances, this resulting information might not matter to you. But if a scan isn't succeeding, more verbose output can help diagnose any issues.

Getting the OS Fingerprint with -O

Another very useful flag is -O, which instructs Nmap to look at the network traffic's OS fingerprint to determine which OS is running on the systems being scanned:

```
# nmap -O 10.0.1.5
```

Just as a human has a unique set of fingerprints, an operating system has characteristics that, taken in combination, are unique to the OS and often point to a specific version or even an OS patch level. But this information isn't guaranteed to be correct. For instance, custom network stacks can throw off OS detection by changing the fingerprint, and even marginally skilled coders could deliberately shape the network traffic from a system to make it look like it has a different OS. Regardless, OS fingerprinting is a highly useful data point in your asset database.

Listing 8-3 shows example output of the -O flag.

```
Nmap scan report for 10.0.1.5
Host is up (0.0035s latency).
Not shown: 996 filtered ports
PORT     STATE SERVICE
135/tcp  open  msrpc
139/tcp  open  netbios-ssn
445/tcp  open  microsoft-ds
5357/tcp open  wsdapi
MAC Address: 70:85:C2:4A:A9:90 (ASRock Incorporation)
```
❶ Warning: OSScan results may be unreliable because we could not find at least 1 open and 1 closed port
```
Device type: general pupose|phone|specialized
```
❷ Running (JUST GUESSING): Microsoft Windows Vista|2008|7|Phone|2012 (93%), FreeBSD 6.X (86%)
```
OS CPE: cpe:/o:microsoft:windows_vista::- cpe:/o:microsoft:windows_vista::sp1
cpe:/o:microsoft:windows_server_2008::sp1 cpe:/o:microsoft:windows_7
cpe:/o:microsoft:windows cpe:/o:microsoft:windows_8
cpe:/o:microsoft:windows_server_2012 cpe:/o:freebsd:freebsd:6.2
```
❸ Aggressive OS guesses: Microsoft Windows Vista SP0 or SP1, Windows Server 2008 SP1, or Windows 7 (93%), Microsoft Windows Vista SP2, Windows 7 SP1, or Windows Server 2008 (93%), Microsoft Windows Phone 7.5 or 8.0 (92%), Windows Server 2008 R2 (92%), Microsoft Windows 7 Professional or Windows 8 (92%), Microsoft Windows Embedded Standard 7 (91%), Microsoft Windows Server 2008 SP1 (91%), Microsoft Windows Server 2008 R2 (90%), Microsoft Windows 7 (89%), Microsoft Windows 8 Enterprise (89%)
```
No exact OS matches for host (test conditions non-ideal).
Network Distance: 1 hop
```

Listing 8-3: nmap output with OS fingerprinting

If Nmap is uncertain about its fingerprinting, it will clearly report that ❶. But Nmap tells you its best guesses for the running OS ❷, and it provides *common platform enumeration* (*CPE*; a standardized reference to specific OS and software packages) references for those guesses. It will even tell you it's making a more aggressive guess ❸ where it does its best to figure out exactly which version of Windows is on that host. In this case, this system is running Windows 10.

Making Nmap "Aggressive"

The "aggressive" flag -A combines the OS fingerprinting option with version detection and script scanning (equivalent to the flag combination -O -sV --script=default --traceroute). It provides even more information about the host. This scan can be intrusive, and system owners are likely to consider it hostile if they are the target of an aggressive scan. Listing 8-4 shows example output of the -A flag on the same host (10.0.1.5).

```
# nmap -A 10.0.1.5
--snip--
Nmap scan report for 10.0.1.5
Host is up (0.0035s latency).
Not shown: 996 filtered ports
PORT     STATE SERVICE      VERSION
135/tcp  open  msrpc        Microsoft Windows RPC
139/tcp  open  netbios-ssn  Microsoft Windows 98 netbios-ssn
445/tcp  open  microsoft-ds Microsoft Windows 7 or 10 microsoft-ds
5357/tcp open  http         Microsoft HTTPAPI httpd 2.0 (SSDP/UPnP)
❶ |_http-server-header: Microsoft-HTTPAPI/2.0
|_http-title: Service Unavailable
MAC Address: 70:85:C2:4A:A9:90 (Unknown)
--snip--
Host script results:
|_nbstat: NetBIOS name: GAMING-PC, NetBIOS user: <unknown>,
NetBIOS MAC: d8:cb:8a:17:99:80 (Micro-star Intl)
| smb-os-discovery:
❷ |   OS: Windows 10 Home 10586 (Windows 10 Home 6.3)
|   OS CPE: cpe:/o:microsoft:windows_10::-
|   NetBIOS computer name: GAMING-PC
|   Workgroup: WORKGROUP
|_  System time: 2020-05-01T16:44:35-04:00
| smb-security-mode:
|   account_used: guest
|   authentication_level: user
|   challenge_response: supported
|_  message_signing: disabled (dangerous, but default)
|_smbv2-enabled: Server supports SMBv2 protocol

TRACEROUTE
HOP RTT     ADDRESS
1   3.53 ms 10.0.1.5
```

Listing 8-4: Aggressive scanning nmap output

In this case, the more aggressive scanning (NetBIOS checks) determined the actual version of Windows running on the host ❷. The aggressive scan can also determine additional information about the HTTP server running on port 5357 ❶.

Modifying the Output Format with -o

Of particular importance for this guide is the -o flag, which lets you output in default format (-oN), XML (-oX), or greppable (-oG) text. You can also use the -oS flag to output in the "script kiddie" format, but that is a novelty format and unlikely to be useful to you in this book!

By using the XML flag, you can output your scan results in a format that is easily parseable by XML-aware Python scripts. Let's scan the same host we did previously but output in XML format using this command:

```
# nmap -oX output.xml 10.0.1.5
```

The Nmap tool generates some basic output to the screen. But the actual XML output is in *output.xml*, so look inside that file as shown in Listing 8-5.

```
# cat output.xml❶ | xmllint --format -❷
❸ <?xml version="1.0" encoding="UTF-8"?>
<!DOCTYPE nmaprun>
<?xml-stylesheet href="file:///usr/bin/../share/nmap/nmap.xsl" type="text/
xsl"?>
<!-- Nmap 7.01 scan initiated Sat Apr 4 09:26:56 2020 as: nmap -oX output.xml
10.0.1.5 -->
❹ <nmaprun scanner="nmap" args="nmap -oX output.xml 10.0.1.48"
start="1523118416" startstr="Sat Apr 4 09:26:56 2020" version="7.01"
xmloutputversion="1.04">
❺  <scaninfo type="syn" protocol="tcp" numservices="1000"
    services="1,3-4,6-7,9,13,17,19-26,
  --snip--
      64623,64680,65000,65129,65389"/>
    <verbose level="0"/>
    <debugging level="0"/>
    <host starttime="1523118416" endtime="1523118436">
      <status state="up" reason="arp-response" reason_ttl="0"/>
❻    <address addr="10.0.1.48" addrtype="ipv4"/>
❼    <address addr="70:85:C2:4A:A9:90" addrtype="mac"/>
❽    <hostnames>
      </hostnames>
      <ports>
        <extraports state="filtered" count="996">
          <extrareasons reason="no-responses" count="996"/>
        </extraports>
❾      <port protocol="tcp" portid="135">
          <state state="open" reason="syn-ack" reason_ttl="128"/>
          <service name="msrpc" method="table" conf="3"/>
        </port>
--snip--
```

```
            </ports>
            <times srtt="1867" rttvar="254" to="100000"/>
        </host>
❿  <runstats>
            <finished time="1523118436" timestr="Sat Apr 4 09:27:16 2020"
elapsed="20.14" summary="Nmap done at Sat Apr 4 09:27:16 2020; 1 IP address (1
host up) scanned in 20.14 seconds" exit="success"/>
            <hosts up="1" down="0" total="1"/>
        </runstats>
</nmaprun>
```

Listing 8-5: An Nmap scan output in XML format

Take a look inside the file using the cat command ❶. The xmllint command ❷ formats the output to indent it properly and make it more readable. (The trailing - in the xmllint command instructs it to take its input from STDIN, allowing us to pipe the output of the previous command into it.)

The first few lines ❸ make up the header, which you don't need to worry about. Next, you see some basic information about the scan, including the command line that generates this specific output and when it ran ❹. Then there are more scan parameters, including the specific ports that Nmap checked ❺. You see the IP ❻ and MAC ❼ addresses, as well as any hostnames that Nmap detected for this host ❽. In this scan, there are no hostnames, but if there were, the output would look something like this:

```
<hostnames>
        <hostname name="scanme.nmap.org" type="user"/>
        <hostname name="scanme.nmap.org" type="PTR"/>
</hostnames>
```

Next, you see some very interesting output—the details of each open port that Nmap found. TCP port 135 ❾ is open. You know it's open because you received a SYN-ACK packet after probing it, the time-to-live (TTL) of packets coming from it is 128 hops, and it's running the Microsoft Remote Procedure Call (MSRPC) protocol. The output also contains the overall run statistics, including the number of hosts found and how long it took to run the scan ❿.

Customize It

I highly recommend experimenting with the Nmap options for your scan to find a set that work best in your environment: that means producing useful information while not overloading the network or causing issues on the systems you're scanning. Here are a few places to start:

- **Different scan types:** Instead of the default SYN scan, try others.
- **Scan speed:** Limit your scan activity to keep your network unclogged, especially over a low-bandwidth connection.
- **Different OS-fingerprinting options:** Some options are more effective than others, depending on the devices you scan and your overall network configuration.

Zenmap, a graphical frontend for Nmap, is useful if you plan to do a lot of different Nmap scans and want a simpler method than repeatedly creating long shell commands. Zenmap provides output in a browsable format that can make digesting scan results easier.

Getting Started with OpenVAS

In this section, you'll become comfortable running scans with OpenVAS from the web GUI (Greenbone Security Assistant) and the command line. OpenVAS isn't a very user-friendly tool, so it's important you're familiar with its options before you start running scans and analyzing the resulting data.

This discussion won't be a comprehensive OpenVAS tutorial. But it will give you enough familiarity with the web and command line interfaces to generate XML scan results you can use in your vulnerability management system.

Running a Basic OpenVAS Scan with the Web GUI

Here you'll learn how to run a basic OpenVAS scan from the Greenbone web GUI, and in the next section, you'll run a scan from your command line.

Log into Greenbone at *https://localhost:4000/*. The default username is **admin**, and the password is **admin**, but you've changed those credentials already, haven't you? Because Greenbone uses a self-signed *transport layer security (TLS)* certificate, your browser will probably warn you that the site isn't trusted. This is expected; just click through and continue to the login page. When you successfully log in, you should see an empty Greenbone dashboard: you'll fill it with information as you run scans and discover information about your environment (see Figure 8-1).

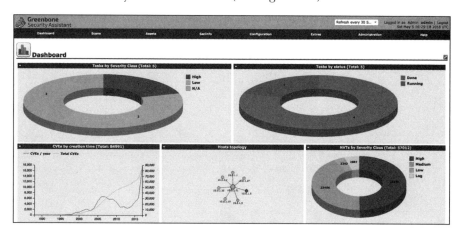

Figure 8-1: A Greenbone dashboard after you run a scan

You'll spend most of your time in the Scans tab where you can run scans and look at the results.

Setting Up Targets and Configurations

You can set up the scan tasks in the Tasks tab, but first you need to set up targets and scan configurations on the Configuration tab.

At the upper left of the Targets page on the Configuration tab is a small star icon; this is Greenbone's indicator for creating a new item (see Figure 8-2). Click the **star** icon to create a new target.

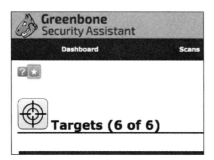

Figure 8-2: Creating a new target

Once you click the star, the New Target configuration page will appear (see Figure 8-3).

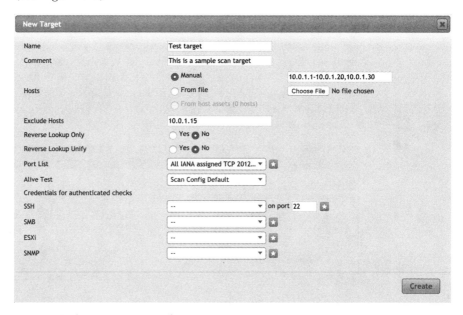

Figure 8-3: The New Target configuration page

The configuration window contains lots of options, but for now, let's focus on the basics. Name the target and, if you like, add a description. The most important option is the Hosts field: enter the hosts you want to scan, separating each host with a comma. You can specify hosts using any of the following methods, alone or in combination:

- IP address
- CIDR IP range
- Dash range (for example, *10.0.1.1-10.0.1.3* refers to 10.0.1.1, 10.0.1.2, and 10.0.1.3)

Just below this field is the option to exclude one or more hosts; using the same format as for the Hosts field, exclude any IPs that you don't want in the scan. For now, we'll leave the rest of the page at the defaults. Click the **Create** button.

Under Configuration ▶ Scan Configs, you'll see a list of different built-in scan configurations (see Figure 8-4).

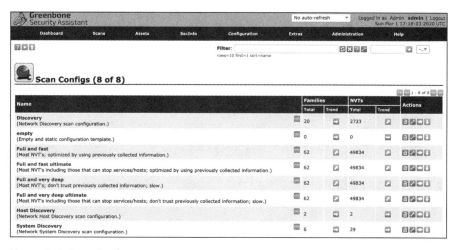

Figure 8-4: Scan Configs

You'll likely want to customize one of these scan types for your own environment. But we'll stick to using **Full and fast** for now; it's a good compromise between speed and thoroughness.

Creating a Task

Now that you have a target set and a scan config to use, you can create the scan task. Return to **Scans ▶ Tasks** and click the **star** icon in the upper left to create a new task. This will open a configuration menu, like the one shown in Figure 8-5.

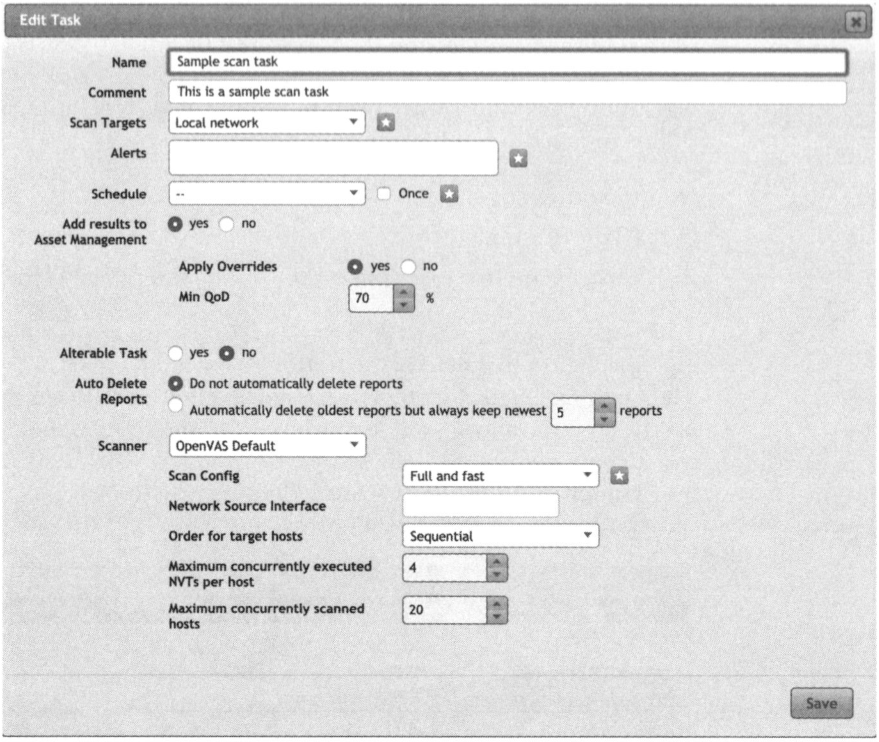

Figure 8-5: New scan task

Much like the target configuration, I recommend you stick to minimal customizations for now, as in Figure 8-5. Name the task and choose the scan target you just created. Scroll down the configuration window, select the **Full and fast** scan config, and click **Create**.

You now have a shiny new scanning task. You can prompt OpenVAS to start the task by clicking the **play** icon on the right of the task list (see Figure 8-6).

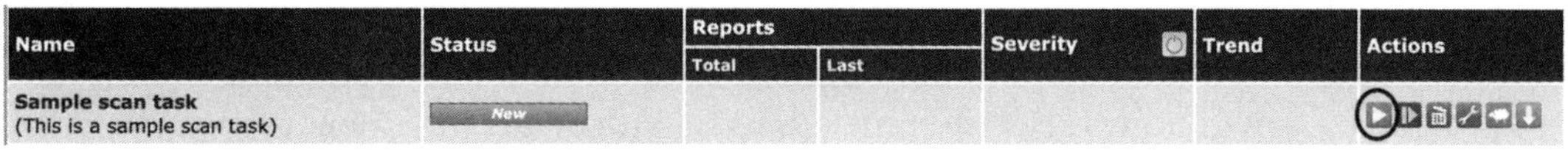

Figure 8-6: Your new task

The **Scans ▸ Tasks** window will list all your tasks and provide some information about each task, such as the severity of the vulnerabilities discovered by the task and the last time the task was completed (see Figure 8-7).

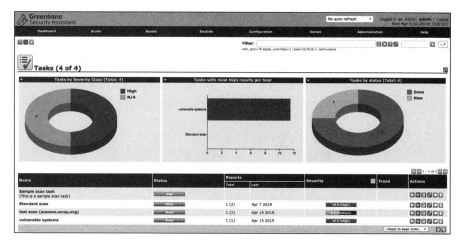

Figure 8-7: The **Scans ▸ Tasks** window after you add a few tasks

Now that your scan has started, you can watch its progress in the **Scans ▸ Reports** page. Even though the scan has just begun, there's a good chance there's already some information there. The scan might take several hours to complete, depending on the number of hosts you're scanning.

Exporting Your Scan Reports

Once the report status indicates Done, the scan is complete, and you can export the report. We'll ingest the XML-formatted report into our database in the next chapter, but first we need to generate the report.

The easiest way to do this with Greenbone is to export the report from the **Scans ▸ Reports** page. Click the report you want to export (in the Date column), ensure that the drop-down box next to this icon is set to **XML** or **Anonymous XML**, and then click the **down-pointing arrow** icon to download the report (see Figure 8-8).

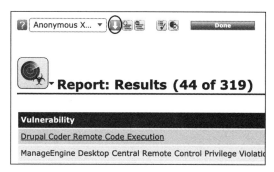

Figure 8-8: Exporting a report

You can look at the resulting file in a text or XML editor.

Running a Basic Scan from the Command Line

Now you know how to create OpenVAS targets, tasks, and scan reports through the Greenbone web GUI. But to integrate automated scanning into the rest of your vulnerability management program, you'll need to use the omp command line tool, which was installed with OpenVAS. Then you can schedule your scans with the same scripts you use to ingest the scan outputs into your vulnerability database (see Chapter 9).

To get omp to work, you need to modify the init scripts for openvas -manager and openvas-gsd. The reason is that the packaged version of OpenVAS is configured to listen for commands on a Unix socket (not a network port). Unfortunately, omp can only communicate on a network port, so you must change this configuration. To do so, find the DAEMON_ARGS variable in /etc/init.d/openvas-manager and modify it to read DAEMON_ARGS="-a 127.0.0.1 -p 9390".

Next, in /etc/init.d/openvas-gsa, modify the DAEMON_ARGS line to read DAEMON_ARGS="--mlisten 127.0.0.1 -m 9390".

Then reload the init scripts (systemctl daemon-reload) and restart openvas-manager (service openvas-manager restart) and openvas-gsd (service openvas-gsa restart).

Specify the username and password in ~/omp.config (it's the same as for the web GUI, so use whatever you've changed it to from admin/admin). When you run scheduled jobs, you'll use the -c command line option to point at this config file, as shown in the following code. This ensures that omp finds the correct configuration file when you run it from *cron*, whether or not you run it as the scanuser user.

```
$ omp -c /home/scanuser/omp.config -X "<help></help>"
```

Now that your OpenVAS command line tool is up and ready to go, let's test it by setting up some scans.

Testing the Command Line Tool

OpenVAS uses XML for input and output, meaning you'll send commands as XML blocks to omp using the --xml or -x flag.

Test omp by sending a simple command and verifying that the result looks like Listing 8-6.

```
$ omp -X "<help></help>"
<help_response status_text="OK" status="200">
    AUTHENTICATE           Authenticate with the manager.
    COMMANDS               Run a list of commands.
    CREATE_AGENT           Create an agent.
    CREATE_ALERT           Create an alert.
    CREATE_ASSET           Create an asset.
    CREATE_CONFIG          Create a config.
    CREATE_CREDENTIAL      Create a credential.
```

```
--snip--
    VERIFY_REPORT_FORMAT    Verify a report format.
    VERIFY_SCANNER          Verify a scanner.
</help_response>
```

Listing 8-6: Testing omp

If you get a result other than status="200" in the first response line, it's likely that your credentials are incorrect. Verify those, and if you still experience issues, check back to the OpenVAS setup steps in Chapter 7 and ensure you've followed the steps in "Updating OpenVAS."

Creating a Scan Task with omp

Because you've already created a scan target and scan configuration via Greenbone, you can use them to initiate scans from the command line. It's a lot easier to build and test scan targets and configurations using the web GUI and then schedule scans and export reports at the command line, using XML as little as possible.

NOTE   *You can find a full XML command reference at* http://docs.greenbone.net/API /OMP/omp-7.0.html.

To check our list of available targets and configs, use omp with the -T and -g flags, as in Listing 8-7.

```
$ omp -T
❶ c8f84568-94ea-4528-b049-56f4029c1368    Target for immediate scan of IP
  10.0.1.0/24
  7fc8000a-28f7-45ea-bd62-9dec89a1f679    Target for immediate scan of IP 10.0.1.1
  a6f26bd5-f1e3-4fd3-88fc-8aa65dd487bc    Target for immediate scan of IP 10.0.1.7
  206a5a14-ab30-462c-b191-440a30daeb17    Target for immediate scan of IP 10.0.1.8
  6539fd3c-871c-43ff-be9c-9768e6bebddd    test target
$ omp -g
❷ 8715c877-47a0-438d-98a3-27c7a6ab2196    Discovery
  085569ce-73ed-11df-83c3-002264764cea    empty
  daba56c8-73ec-11df-a475-002264764cea    Full and fast
  698f691e-7489-11df-9d8c-002264764cea    Full and fast ultimate
  708f25c4-7489-11df-8094-002264764cea    Full and very deep
  74db13d6-7489-11df-91b9-002264764cea    Full and very deep ultimate
  2d3f051c-55ba-11e3-bf43-406186ea4fc5    Host Discovery
  bbca7412-a950-11e3-9109-406186ea4fc5    System Discovery
```

Listing 8-7: Getting target and scan config IDs

The omp -T flag returns a list of configured scan targets ❶ with *universally unique identifiers (UUIDs)*. The omp -g flag returns a list of configured scan types ❷, again with matching UUIDs. With these target and config IDs, create a scan task with the -C flag, as shown in Listing 8-8.

```
$ omp -C --target=c8f84568-94ea-4528-b049-56f4029c1368 --config=698f691e-7489-
11df-9d8c-002264764cea --name="Scan Task"
❶ dd3617ce-868f-457a-a2f8-bfb7bdb1b8ff
```

Listing 8-8: Creating a task and getting its ID

You can also do this with an XML command, but the -C format is much simpler! The response you'll get is a UUID that refers to an OpenVAS task. Think of the task as a combination of a scan configuration and a target; you can kick off that task on demand by referring to its UUID ❶, as shown in Listing 8-9.

```
$ omp --start-task dd3617ce-868f-457a-a2f8-bfb7bdb1b8ff
<start_task_response status_text="OK, request submitted" status="202"><report
_id>4dc106f1-cf3a-47a1-8a71-06b25d8d2c70</report_id></start_task_response>
```

Listing 8-9: Starting the task

When the task is complete, you'll need the report_id to look at the results. The easiest way to tell if the scan is running is to look at the **Tasks** page in Greenbone. You should see a task in progress in the table at the bottom of the page (see Figure 8-9).

Figure 8-9: A running scan task

You can also check on the scan task's progress by using the --get_tasks command, which will give you an exhaustive rundown of the scan's current status. Listing 8-10 shows a (truncated!) example.

```
$ omp --get-tasks dd3617ce-868f-457a-a2f8-bfb7bdb1b8ff|xmllint --format -
<?xml version="1.0"?>
<get_tasks_response status_text="OK" status="200">
  <apply_overrides>0</apply_overrides>
  <task id="dd3617ce-868f-457a-a2f8-bfb7bdb1b8ff">
--snip--
❶ <status>Running</status>
--snip--
</get_tasks_response>
```

Listing 8-10: omp get_tasks output

When the status XML value ❶ changes from Running to Done, the scan is finished. As in Listing 8-5, xmllint formats the output XML into a more readable layout.

Once the scan is complete, you can get the report in XML format using omp --get-report, as shown in Listing 8-11.

```
$ omp --get-report 4dc106f1-cf3a-47a1-8a71-06b25d8d2c70|xmllint --format - > output.xml
```

Listing 8-11: Exporting the report

In Listing 8-11, we redirect the output to a file using >. You can run this command at any time during the scan, but the report will be incomplete. There's a field in the output XML file called scan_run_status; you'll know you have a full report when the value of this field changes from Running to Done. We'll look at this XML output format in more detail in Chapter 9.

Customize It

If there's more than one security analyst in your organization, you can set up additional accounts or integrate with LDAP/RADIUS authentication in the Greenbone web interface to allow others to set up scan targets, run scans, and look at results.

You can do more in-depth scans by configuring SSH and other credentials: this permits OpenVAS to log into some services and conduct deeper scans. You can save these credentials most easily via the Greenbone web GUI under the Configuration tab.

Getting Started with cve-search

Although we'll use cve-search primarily for its comprehensive database of CVE information, it's very useful on its own for finding detailed information about any vulnerability with a CVE ID.

Searching for CVE IDs

We'll use *search.py*, which is in *cve-search/bin/* (see the section "Setting Up cve-search" in Chapter 7), to search for CVEs that affect a product (identified by its CPE) and search for a specific CVE.

In Listing 8-12, we use the -p (for product) flag to find all CVEs that affect Windows 10.

```
$ ./search.py -p o:microsoft:windows_10
```

Listing 8-12: Searching for CVEs that affect Windows 10

This command outputs a huge result set, including full details of all CVEs found. In Listing 8-13, I've restricted the results returned to CVE ID by passing the value cveid to the -o flag.

```
$ ./search.py -p o:microsoft:windows_10 -o cveid
CVE-2015-6184
CVE-2015-6051
CVE-2015-6048
CVE-2015-6042
CVE-2016-1002
CVE-2016-1005
```

```
CVE-2016-1001
CVE-2016-1000
CVE-2016-0999
CVE-2016-0998
CVE-2016-0997
CVE-2016-0996
--snip--
```

Listing 8-13: CVE IDs that affect Windows 10

The vulnerability management system you're building will handle tasks like this entirely with the database. But the commands like the one in Listing 8-12 can be useful for filtering a list of vulnerabilities to just those that affect the operating systems your organization has deployed.

Finding Out More About a CVE

To find out more about a specific CVE, specify a CVE ID using the -c flag. Searching for CVE-2016-0996 from Listing 8-13 gives us the results in Listing 8-14.

```
$ ./search.py -c CVE-2016-0996 -o json | python -m json.tool❶
{
    "Modified": "2016-03-16T13:53:26.727-04:00",
    "Published": "2016-03-12T10:59:16.853-05:00",
 ❷ "access": {
        "authentication": "NONE",
        "complexity": "MEDIUM",
        "vector": "NETWORK"
    },
 ❸ "cvss": 9.3,
    "cvss-time": "2016-03-16T09:46:38.087-04:00",
    "id": "CVE-2016-0996",
 ❹ "impact": {
        "availability": "COMPLETE",
        "confidentiality": "COMPLETE",
        "integrity": "COMPLETE"
    },
 ❺ "references": [
        "https://helpx.adobe.com/security/products/flash-player/apsb16-08
        .html",
        "http://www.zerodayinitiative.com/advisories/ZDI-16-193/"
    ],
 ❻ "summary": "Use-after-free vulnerability in the setInterval method in
    Adobe Flash Player before 18.0.0.333 and 19.x through 21.x before
    21.0.0.182 on Windows and OS X and before 11.2.202.577 on Linux, Adobe
    AIR before 21.0.0.176, Adobe AIR SDK before 21.0.0.176, and Adobe AIR
    SDK & Compiler before 21.0.0.176 allows attackers to execute arbitrary
    code via crafted arguments, a different vulnerability than CVE-2016-0987,
    CVE-2016-0988, CVE-2016-0990, CVE-2016-0991, CVE-2016-0994, CVE-2016-0995,
    CVE-2016-0997, CVE-2016-0998, CVE-2016-0999, and CVE-2016-1000.",
```

```
❼ "vulnerable_configuration": [
      "cpe:2.3:a:adobe:flash_player_esr:18.0.0.329",
      "cpe:2.3:o:microsoft:windows",
      "cpe:2.3:o:apple:mac_os_x",
--snip--
```

Listing 8-14: Details of CVE-2016-0996

The output contains the access vectors for this vulnerability ❷ and all the relevant CVSS and common weakness enumeration (CWE) information. (Note that, like OpenVAS, cve-search doesn't provide CVSSv3 scores, but only a CVSSv2 score ❸ and details, via its access ❷ and impact ❹ sections.) Another section lists all external references ❺ for this vulnerability: they include a US-CERT alert, the Microsoft patch details, and a number of third-party notifications as well. There's a human-readable summary of the vulnerability ❻ and CPE information for all the known systems (OS or application) that are subject to this vulnerability ❼.

By using the -o json flag, the command returns a block of JSON without newlines or indentation, so you'll need to pipe the output through the Python tool json.tool ❶ for readable formatting.

Text Searching the CVE Database

We can also use the -f flag to do an arbitrary text search across the CVE database summary fields. Listing 8-15 shows the first (of many) results of a search for "buffer overflow."

```
$ ./search.py -f "buffer overflow"
{
  "Modified": "2008-09-05T16:40:25.38-04:00",
  "Published": "2005-01-10T00:00:00.000-05:00",
  "_id": {
    "$oid": "5706df571d41c81f2d58a882"
  },
  "access": {
    "authentication": "NONE",
    "complexity": "HIGH",
    "vector": "NETWORK"
  },
  "cvss": 5.1,
  "cvss-time": "2004-01-01T00:00:00.000-05:00",
  "id": "CVE-2004-1112",

--snip--
```

Listing 8-15: Searching for CVEs related to buffer overflows

With this search tool, you can find complete information for any vulnerabilities matching your specific criteria. By combining flags, you can compile lists of, for instance, buffer overflow vulnerabilities from the last year that affect Linux.

Customize It

Although *search.py* doesn't currently support output formatting for CVE searches (only for product searches), you can use Linux command line utilities, like jq, to parse JSON strings and even pull out specific fields. Use these tools to write bash scripts to do whatever data manipulation you please.

Here are a couple of examples:

- Generate a weekly list of relevant (for instance, high severity and affecting Windows 10) vulnerabilities that have been added to the database and then email it to security staff.
- Regularly look at "interesting" CVEs (perhaps one that is affecting your organization that you've not yet fully addressed) and note whether new reference URLs have been added to their records.

Your full vulnerability management system and some custom scripting can handle the preceding two use cases, but sometimes it's easier and more straightforward to perform the same task with command line tools.

Summary

In this chapter, you learned about the main data sources—Nmap, OpenVAS, and cve-search—for your vulnerability management program. You tried each tool to understand the types of information you can gather and started thinking about how to use the three tools to provide a more complete vulnerability picture of your organization.

In Chapter 9, you'll take the first steps toward painting that picture by writing scripts that will parse and import scanner outputs into your vulnerability database.

9

CREATING AN ASSET AND VULNERABILITY DATABASE

In this chapter, you'll learn to get the outputs from OpenVAS and Nmap into your Mongo database. You'll start by exploring some general practices for working with data. Next, you'll look at the XML output formats of both tools and learn to select the specific data fields you're interested in. Then you'll walk through a few Python scripts that will collect all this data, generate Mongo documents, and insert those documents into the database.

Preparing the Database

To design a database, you'll need to understand your desired outcomes and the analysis you'll have to do to achieve those outcomes. Then you can think about the data you'll have to collect and the model you need to make those analyses possible.

In this situation, you want a more secure environment to improve your vulnerability posture. To make that improvement, you need information about the hosts you're trying to secure, and that information comes in two varieties: *persistent* and *dynamic*.

Persistent information doesn't change (or rarely changes); dynamic information changes frequently. Your environment dictates how a data point is categorized. For example, in some networks, IP addresses are statically assigned, but in others, *dynamic host configuration protocol (DHCP)* might assign a different address after every reboot or even daily. Persistent data will be collected once and updated as needed. Dynamic data associated with a device will be updated every time a vulnerability scan is run against that device.

Table 9-1 describes the host-based data that we'll collect from the Nmap and OpenVAS scans and how it's categorized.

Table 9-1: Relevant Host Data

Data type	Notes
Persistent	
Hostname	If available. In some cases, multiple hostnames might be reported.
MAC address	If available.
IP address	IPv4 address. If your environment uses IPv6, you can modify the scripts accordingly to capture this information.
OS / CPE	The OS version that was detected, including CPE if available.
Dynamic	
Vulnerabilities	Includes details reported by OpenVAS and a reference to the cve-search CVE entry.
Ports	Ports that are open (listening for incoming data) including number, protocol, and detected service.
Last scan date	Automatically generated.

In the record (document) for each host, you'll include a *vulnerability identifier* for each vulnerability discovered on that host. You can use this identifier to relate hosts to specific vulnerability and exploit information. A good deal of data related to each vulnerability is stored in its own dataset (*collection*, in Mongo parlance). This means that you'll have a collection of hosts and a collection of vulnerabilities with mappings from the former to the latter.

In other words, vulnerability information is orthogonal to hosts: one host might have one or more vulnerabilities, but each vulnerability is a data item, which you can relate to one or more hosts. The same is true of exploits. Table 9-2 contains the vulnerability data you'll collect in the scripts later in the chapter.

Table 9-2: Relevant Vulnerability Data

Data type	Notes
CVE/BID ID	CVE or Bugtraq ID for the vulnerability. This is an industry standard identifier.
Date reported	When the vulnerability was first reported—by the vendor, by a third party, or by active exploitation.
Affected software	Names and CPEs of software (or OS) affected by the vulnerability.
CVSS	CVSS score for the vulnerability.
Description	Free-form text description of the vulnerability.
Advisory URLs	URLs pointing to advisories about the vulnerability that might contain more information.
Update URLs	URLs pointing to update information for addressing the vulnerability.

Understanding the Database Structure

Although MongoDB can accept unstructured data, your scripting and data analysis will be much easier if you have some idea of the data types you're trying to capture and how you want to structure them in your Mongo documents. If you're using a relational database, such as SQL, this step is absolutely essential, because you can't insert data into an empty database with no structure.

RELATIONAL VS. NON-RELATIONAL DATABASES

A full discussion of the difference between database types could be its own substantial volume and is beyond the scope of this book. But I'll provide a crash course.

A *relational database* is what most people think of when you talk about databases: it's a database that contains tables consisting of rows and columns of structured data. Each row is identified by a unique key. Connections between tables are made by sharing key values, so a value in one table might point to an entire row of data in another table—hence the name "relational."

Databases that don't share this structure are *non-relational databases*. A wide variety of databases exist under this heading, including MongoDB. These non-relational databases might be as simple as a list of key-value pairs, but they can also allow you to arbitrarily structure data.

(continued)

Let's look at an example. A relational database might have a table labeled NAME with columns FirstName, MiddleName, LastName, like so:

FirstName	MiddleName	LastName
Andrew	Philip	Magnusson
Jorge	Luis	Borges

But a rigid structure like this doesn't always make sense: the standard first-middle-last paradigm can be difficult to map onto names in other cultures. For instance, Jorge Luis Borges, the Argentine author, had several more names: *Jorge Francisco Isidoro Luis Borges Acevedo*. In MongoDB, you might have a *collection* (roughly analogous to a table) called NAME that can include all sorts of name structures in their own *documents* (roughly analogous to data rows), for instance:

```
{
    "FirstName":"Jorge",
    "MiddleNames": ["Francisco","Isidoro","Luis"],
    "LastNames":["Borges","Acevedo"]
}
```

A simpler name might just have FirstName and a single value in LastNames:

```
{
    "FirstName":"Alexander",
    "LastNames":"Lovelace"
}
```

Each type of database has its advantages. Relational databases have a predefined data structure that makes queries, indexing, and database maintenance very fast but at a cost: they have a rigid structure that cannot be changed without major effort, especially once the database is being used in a production environment. Non-relational databases let you define your data structure more loosely and change it as you go, offering flexibility. But they'll never be quite as fast, and you can't count on specific data fields existing in your documents. In the preceding NAME data example, you'll need to make sure your code is sufficiently robust to not crash when a MiddleNames field isn't present!

Now that you have an idea of the kinds of data you want to collect, you can start building your database structure. In this section, we'll look at how you might represent this data in Mongo or in SQL formats.

Listing 9-1 shows an example of host data in *JavaScript Object Notation (JSON)* format, which is representative of the MongoDB internal data structure. Strictly speaking, Mongo stores its data in *binary JSON (BSON)*, a more compact way of representing the same data as JSON. But for the purposes of interacting with Mongo, you'll use JSON.

```
{
❶ "_id" : ObjectId("57d734bf1d41c8b71daaee0e"),
❷ "mac" : {
        "vendor" : "Apple",
        "addr" : "6C:70:9F:D0:31:6F"
    },
    "ip" : "10.0.1.1",
❸ "ports" : [
        {
            "state" : "open",
            "port" : "53",
            "proto" : "tcp",
            "service" : "domain"
        },
--snip--
    ],
    "updated" : ISODate("2020-01-05T02:19:11.966Z"),
❹ "hostnames" : [
"airport",
"airport.local"
],
    "os" : [
        {
            "cpe" : [
                "cpe:/o:netbsd:netbsd:5"
            ],
            "osname" : "NetBSD 5.0 - 5.99.5",
            "accuracy" : "100"
        }
    ],
❺ "oids" : [
        {
            "proto" : "tcp",
            "oid" : "1.3.6.1.4.1.25623.1.0.80091",
            "port" : "general"
        }
    ]
}
```

Listing 9-1: An example JSON host description document

NOTE   *In the JSON format, keys and values are delineated by double quotes. A* key *is a unique string that labels the following value. A* value *can be a simple string, a nested JSON document (delineated by curly brackets), or a list of strings or nested documents (surrounded by square brackets). This format lets you build a sophisticated data structure that is easy to parse and traverse.*

In Listing 9-1, the _id field ❶, auto-generated by Mongo, uniquely identifies the document within the database. The value of the mac field ❷ is a nested document that contains the MAC address and the MAC vendor. The ports key ❸ contains a list of documents that each contain information about an open port. Because a host often has different hostnames depending on which one you ask—*domain name system (DNS)* servers might use one name and a NetBIOS lookup something else—hostnames ❹ is a list instead of a single value. The oids key ❺ contains a list of documents containing an OID, a protocol, and a port that OID was detected on. The *OID* is a unique vulnerability identifier generated by OpenVAS that you'll use to map vulnerabilities to hosts. In the vulnerabilities collection, there will be one unique document (representing a specific vulnerability) for each unique OID.

SQL TABLE STRUCTURE

If you're using SQL, you'll need to know the data type you'll be storing for each data field to define your database tables. Here are example table definitions you might use in SQL. Keep in mind that the definitions in Listing 9-2 are not perfectly optimized and there are ways to improve the structure for large databases—see "Customize It" on page 86 for more. (The following definitions are for MySQL; you might need to adjust for other SQL flavors.)

```
❶ CREATE TABLE hosts
      (macid CHAR(17), macvendor VARCHAR(25),
      ip VARCHAR(15), hostname VARCHAR(100),
   ❷ updated DATETIME DEFAULT CURRENT_TIMESTAMP ON UPDATE CURRENT_TIMESTAMP,
      id INT AUTO_INCREMENT PRIMARY KEY);
❸ CREATE TABLE ports
      (id INT AUTO_INCREMENT PRIMARY KEY,
   ❹ host_id INT NOT NULL, state VARCHAR(6),
      port INT, protocol VARCHAR(3), service VARCHAR(25),
   ❺ FOREIGN KEY(host_id) REFERENCES hosts(id));
❻ CREATE TABLE os
      (id INT AUTO_INCREMENT PRIMARY KEY,
      cpe VARCHAR(50), osname VARCHAR (50), accuracy INT,
      FOREIGN KEY(host_id) REFERENCES hosts(id));
❼ CREATE TABLE hostoid
      (id INT AUTO_INCREMENT PRIMARY KEY,
      FOREIGN KEY(oid_id) REFERENCES oids(id),
      FOREIGN KEY(host_id) REFERENCES hosts(id));
```

Listing 9-2: MySQL table definitions for host data

These commands will create tables in an existing SQL database. Because SQL doesn't nest data directly, as you can do in Mongo, you'll need to split your data into multiple tables. Here you have table definitions for hosts ❶, ports ❸, and OS information ❻ and a table mapping hosts to OIDs ❼. You'll need to use keys to bring it all together.

A key in SQL identifies an individual record in a specific table, mostly by using foreign keys in other tables that refer back to the original table. For example, in the ports table is a field (host_id ❹) that is explicitly defined as a foreign key ❺: the id field in the hosts table. This key lets you query the database and find port information for a specific host. The same lines are in the os definition, and all together the three tables are linked to provide direct access to all the persistent host information you need. An updated field ❷ is automatically changed every time a hosts record is changed.

Now let's look at the vulnerability data in Listing 9-3. The JSON for a vulnerability document contains all the information listed in Table 9-2, as well as some extra fields reported by the OpenVAS scanner. If space is at a premium, you might not need to record all this information in your database. But if you have the space, it can't hurt to keep it around for future use.

```
{
    "_id" : ObjectId("57fc2c891d41c805cf22111b"),
  ❶ "oid" : "1.3.6.1.4.1.25623.1.0.105354",
    "summary" : "The remote GSA is prone to a default account authentication
                bypass vulnerability.",
    "cvss" : 10,
    "vuldetect" : "Try to login with default credentials.",
    "solution_type" : "Workaround",
    "qod_type" : "exploit",
    "bid" : "NOBID",
    "threat" : "High",
    "description" : null,
    "proto" : "tcp",
    "insight" : "It was possible to login with default
                credentials: admin/admin",
    "family" : "Default Accounts",
    "solution" : "Change the password.",
    "xref" : "NOXREF",
    "port" : "443",
    "impact" : "This issue may be exploited by a remote attacker to gain
                access to sensitive information or modify system
                configuration.",
  ❷ "cve" : [
        "NOCVE"
    ],
    "name" : "GSA Default Admin Credentials",
    "updated" : ISODate("2016-10-11T00:04:25.596Z"),
    "cvss_base_vector" : "AV:N/AC:L/Au:N/C:C/I:C/A:C"
}
```

Listing 9-3: An example JSON vulnerability description document

As with Listing 9-1, much of this data is fairly self-explanatory, but there are a few important points to note. The oid value ❶ can be added directly to the list of oids in the host document: each vulnerability that OpenVAS finds on a host will be given a separate OID. So the OID will be put into the host document, and the details of the OID will be in the vulnerability collection. When you need to report on vulnerabilities found on a given host, you'll first retrieve the host record and then retrieve the records associated with any OIDs recorded for that host. The cve key ❷ has a list as its value, because individual vulnerabilities are often associated with more than one CVE. In this example, the only CVE reported is NOCVE, which is a standard placeholder when MITRE has not assigned a CVE ID to a vulnerability.

As you proceed through the examples that follow, consider how the scripts build documents in Mongo and make any necessary adjustments for your own needs.

Customize It

Think carefully about your needs and customize the information you collect accordingly. For example, let's say your organization has specific VLANs set up for different purposes. You can add a key-value pair in your Mongo database to specify which VLAN a host is located on, or customize your scripts to determine the VLAN from the host's IP address, to facilitate analysis of which hosts are on which network segments. Depending on your network configuration, this might require consulting external systems or databases.

Return to your data definitions as you think about the new information you're gathering and how you want to store it. If you're using an unstructured database like Mongo, it's easy enough—just add key-value pairs. But if you're using SQL, you'll need to reconfigure your database to define those new data fields.

If you're using SQL, you can optimize the table structure to save some space when handling large datasets. For example, instead of a ports table that maps directly to hosts records using records like port, service, protocol, host ID, you could have a third table, host-to-port mappings, so any one port record could map to any number of hosts, and vice versa. Make a hosts table with host ID and other fields; a ports table with port, service, protocol records; and a host-port table with port ID and host ID records. In small environments, the difference is minimal, but in larger organizations, the space savings could be substantial.

Getting Nmap into the Database

Now you'll start connecting the tools you've deployed. First, you'll need to write a script to input your Nmap scan data into the Mongo database.

Defining the Requirements

As with any script, decide what you want your ingestion script to accomplish: namely, collecting the host data laid out in Table 9-1. We'll start this process by looking at the Nmap XML output to define which portions of it are important.

Listing 9-4 shows a segment of XML output from Nmap (run with the OS detection flag -0).

```
--snip--
<host starttime="1473621609" endtime="1473627403"><status state="up"
reason="arp-response" reason_ttl="0"/>
❶ <address addr="10.0.1.4" addrtype="ipv4"/>
❷ <address addr="B8:E8:56:15:68:20" addrtype="mac" vendor="Apple"/>
❸ <hostnames>
</hostnames>
❹ <ports><extraports state="filtered" count="997">
<extrareasons reason="no-responses" count="997"/>
</extraports>
<port protocol="tcp" portid="22"><state state="open" reason="syn-ack"
reason_ttl="64"/><service name="ssh" method="table" conf="3"/></port>
--snip--
</ports>
❺ <os><portused state="open" proto="tcp" portid="22"/>
<osmatch name="Apple Mac OS X 10.10.2 (Darwin 14.1.0)"
accuracy="100" line="4734">
<osclass type="general purpose" vendor="Apple" osfamily="Mac OS X"
osgen="10.10.X" accuracy="100"><cpe>cpe:/o:apple:mac_os_x:10.10.2</cpe>
</osclass>
</osmatch>
<osmatch name="Apple Mac OS X 10.7.0 (Lion) - 10.10 (Yosemite)
or iOS 4.1 - 8.3 (Darwin 10.0.0 - 14.5.0)" accuracy="100" line="6164">
--snip--
<osclass type="phone" vendor="Apple" osfamily="iOS" osgen="4.X"
❻ accuracy="100"><cpe>cpe:/o:apple:iphone_os:4</cpe></osclass>
<osclass type="phone" vendor="Apple" osfamily="iOS" osgen="5.X"
accuracy="100"><cpe>cpe:/o:apple:iphone_os:5</cpe></osclass>
--snip--
```

Listing 9-4: Excerpt from Nmap XML output

We need to parse out the IP address ❶; MAC address ❷; hostname ❸ (if it's available—it's not here); open ports ❹ with protocol, port number, its state (open or closed), and a guess at the service running on that port; and OS matches ❺. The MAC address also depends on availability: if the destination host is more than one hop from the scanner, the MAC address likely belongs to a router or switch rather than the actual host.

We record all the returned osmatch values ❺ along with corresponding CPE labels and accuracy tags ❻ as a list to reflect Nmap's uncertainty about the match. In this example, multiple CPEs are reported as a 100 percent accuracy match; when you produce a report for this host, you'll have to report all, none, or choose one based on other criteria.

You need to associate all this information with a single host document and distinguish that host document using a field that's present in every scan result. The hostname and MAC addresses might not be present or accurate, so we use the IP address. If the IP changes regularly in your DHCP environment, a Windows NetBIOS name might be a better choice, that is, if you can guarantee continuity, because NetBIOS names must be unique per Windows domain.

You must also decide whether you want to create new host documents for each scan or update existing documents. The reason is that in most use cases, only some of the data changes from scan to scan—most prominently, the list of vulnerabilities. It will save time and effort to update an existing document with the new information you collect.

Building the Script

In Listing 9-5, IP addresses are authoritative, and new data for existing hosts will overwrite old data. Of course, your requirements might differ. The script iterates through an Nmap output XML file and inserts relevant information into a Mongo database.

All the required information for a host is contained within a host tag, so you need a simple loop to find each host tag, pull the appropriate sub-tags, and then do a Mongo document insert for each host into the Mongo hosts database.

```
#!/usr/bin/env python3

❶ from xml.etree.cElementTree import iterparse
  from pymongo import MongoClient
  import datetime, sys

❷ client = MongoClient('mongodb://localhost:27017')
❸ db = client['vulnmgt']

  def usage():
      print ('''
  Usage: $ nmap-insert.py <infile>
      ''')

❹ def main():
      if (len(sys.argv) < 2): # no files
          usage()
          exit(0)

    ❺ infile = open(sys.argv[1], 'r')

    ❻ for event, elem in iterparse(infile):
        ❼ if elem.tag == "host":
              # add some defaults in case these come up empty
              macaddr = {}
              hostnames = []
              os = []
```

```
        addrs = elem.findall("address")
        # all addresses, IPv4, v6 (if exists), MAC
        for addr in addrs:
            type = addr.get("addrtype")
            if (type == "ipv4"):
                ipaddr = addr.get("addr")
            if (type == "mac"): # there are two useful things to get here
                macaddr = {"addr": addr.get("addr"),
                           "vendor": addr.get("vendor")}

        hostlist = elem.findall("hostname")
        for host in hostlist:
            hostnames += [{"name": host.get("name"),
                           "type": host.get("type")}]

        # OS detection
        # We will be conservative and put it all in there.
        oslist = elem.find("os").findall("osmatch")
        for oseach in oslist:
            cpelist = []
            for cpe in oseach.findall("osclass"):
                cpelist += {cpe.findtext("cpe")}
            os += [{"osname": oseach.get("name"),
                    "accuracy": oseach.get("accuracy"),
                    "cpe": cpelist}]

        portlist = elem.find("ports").findall("port")
        ports = []
        for port in portlist:
            ports += [{"proto": port.get("protocol"),
                       "port": port.get("portid"),
                       "state": port.find("state").get("state"),
                       "service": port.find("service").get("name")
                      }]
        elem.clear()

❽ host = {"ip": ipaddr,
          "hostnames": hostnames,
          "mac": macaddr,
          "ports": ports,
          "os": os,
          "updated": datetime.datetime.utcnow()
         }

❾ if db.hosts.count({'ip': ipaddr}) > 0:
        db.hosts.update_one(
               {"ip": ipaddr},
               {"$set": host}
               )
    else:
        db.hosts.insert(host)
```

```
⓿ infile.close() # We're done.

main()
```

Listing 9-5: The nmap-insert.py *code listing for Nmap database insertion*

We import ❶ iterparse from the xml library for the XML parsing,
MongoClient from pymongo to interact with the database, and datetime and sys
for generating the current date and file read/writes, respectively. Fill in
your Mongo server IP ❷ and database information.

We encapsulate the main logic in a main() function ❹ that we call at the
end of the script listing. This function first opens the input file ❺, which is
passed as an argument to the script, loops through every XML element ❻,
and gathers details from each host element ❼. Then it inserts or updates a
Mongo document ❾ with that information for each host ❽. The script takes
the IP address as the canonical identifier of the host, creates a new document if the IP doesn't exist yet, and updates an existing document if that IP
is associated with it. Once the script runs out of XML to parse, it closes the
input file and exits ❿.

Customize It

If your organization uses IPv6, or if IPv6 has finally taken over when you're
reading this, record the IPv6 address instead of ignoring it as Listing 9-5
does. Keep in mind that IPv6 is no more authoritative than IPv4; a single
host might have multiple (and changing!) IPv6 addresses.

You can modify the script in Listing 9-5 to capture more (or all) of the
Nmap output for the database. Say you want to track Nmap's run statistics.
Listing 9-6 shows a Python snippet example to parse the runstats XML
block, which for clarity you might place before the main if elem.tag ==
"host": ❼ statement.

```
if elem.tag == "runstats":
    finished = elem.find("finished")
    hosts = elem.find("hosts")
    elapsed = finished.get("elapsed")
    summary = finished.get("summary")
    hostsUp = hosts.get("up")
    hostsDown = hosts.get("down")
    hostsTotal = hosts.get("total")
```

Listing 9-6: Python snippet to parse runstats *block*

You can also add your own key-value pairs to the Mongo document. For
example, a high-value host tag will help prioritize vulnerabilities, because
vulnerabilities on high-value hosts are more pressing than those on normal
or low-value systems.

You can write one script to collect the data and insert it into the database. Although this is more complicated to build, it simplifies scheduling

because you only need to run one script rather than several. (Hint: use the -oX - flags to send the Nmap output to STDOUT, which you can then pipe into the input to the insertion script.)

Instead of manually parsing the XML, you can use libraries. Two Python modules are available to control Nmap and parse its results: *python-nmap* and *python-libnmap*. Try both to see which you prefer.

BUILD OR BORROW?

A well-known dilemma you might face is whether you should write your own code or use existing Python modules to do much of the work for you. Using existing modules eliminates a lot of manual work, but it leads to more software packages to think about and keep up-to-date. In this script, I chose manual coding and more control, but both options are perfectly legitimate.

Getting OpenVAS into the Database

Once you've extracted the relevant Nmap data and inserted it into the database, the next step is to do the same for OpenVAS.

Defining the Requirements

OpenVAS formats its output by result, which for OpenVAS means any findings, from service detection to specific vulnerabilities. Listing 9-7 shows a result for CVE-2016-2183 and CVE-2016-6329 on host 10.0.1.21.

```
--snip--
<result id="a3e8107e-0e6c-49b0-998b-739ef8ae0949">
❶   <name>SSL/TLS: Report Vulnerable Cipher Suites for HTTPS</name>
    <comment/>
    <creation_time>2017-12-29T19:06:23Z</creation_time>
    <modification_time>2017-12-29T19:06:23Z</modification_time>
    <user_tags>
      <count>0</count>
    </user_tags>
❷   <Host>10.0.1.21<asset asset_id="5b8d8ed0-e0b1-42e0-
    b164-d464bc779941"/></host>
❸   <port>4000/tcp</port>
    <nvt oid="1.3.6.1.4.1.25623.1.0.108031">
      <type>nvt</type>
❹     <name>SSL/TLS: Report Vulnerable Cipher Suites for HTTPS</name>
      <family>SSL and TLS</family>
❺     <cvss_base>5.0</cvss_base>
❻     <cve>CVE-2016-2183, CVE-2016-6329</cve>
❼     <bid>NOBID</bid>
❽     <xref>URL:https://bettercrypto.org/, URL:https://mozilla.github.io
      /server-side-tls/ssl-config-generator/, URL:https://sweet32.info/
      </xref>
```

```
   ❾ <tags>cvss_base_vector=AV:N/AC:L/Au:N/C:P/I:N/A:N|summary=This
     routine reports all SSL/TLS cipher suites accepted by a
     service where attack vectors exists only on HTTPS services.
     |solution=The configuration of this service should be changed so
     that it does not accept the listed cipher suites anymore.

Please see the references for more resources supporting you with this task.
|insight=These rules are applied for the evaluation of the vulnerable
cipher suites:

  - 64-bit block cipher 3DES vulnerable to the SWEET32 attack (CVE-2016-2183).
  |affected=Services accepting vulnerable SSL/TLS cipher suites via HTTPS.
  |solution_type=Mitigation|qod_type=remote_app</tags>
      ❿ <cert>
          <cert_ref id="CB-K17/1980" type="CERT-Bund"/>
          <cert_ref id="CB-K17/1871" type="CERT-Bund"/>
          <cert_ref id="CB-K17/1803" type="CERT-Bund"/>
          <cert_ref id="CB-K17/1753" type="CERT-Bund"/>
--snip--
      </result>
--snip--
</results>
```

Listing 9-7: Example result block from an OpenVAS XML scan report

So what information do you want from this scan? Recall that Table 9-2
identified relevant vulnerability data: cve ❻, bid ❼, date reported, affected
software, CVSS ❺, description ❶❹❾, advisory URLs ❽❿, and update
URLs. OpenVAS reports most of this information, as well as the host ❷
and port ❸ associated with the finding.

The cert section ❿ includes links to known computer emergency
response team (CERT) advisories. Although the sample script in
Listing 9-8 ignores this section, parse this data if it's important to you.

Mapping Vulnerabilities to Hosts

Most important is how you'll structure all the data. The two different sets
of data, vulnerabilities and hosts, have implicit mappings between them:
host A has vulnerabilities X, Y, Z. Vulnerability X is on hosts A, B, C. There
are two obvious ways to represent this mapping. Each host can have a list of
vulnerabilities it's subject to: host A would have a vulnerabilities list of X, Y,
Z within its structure. Alternatively, you could use the same mapping on the
vulnerability side. Vulnerability X would have a host tag that contains the
list of hosts A, B, C.

Both options are valid, and both are one-sided. If you store the data
with the vulnerabilities, host reporting is difficult: you have to look for all
the places that host A appears across the entire vulnerability database. The
reverse is true if you store all the vulnerability IDs with the hosts. In addition, if you store both mappings in both places, you risk ending up with
stale or orphaned mappings. Choose one depending on whether you want
to more easily report on vulnerabilities (and which hosts are affected) or
hosts (and which vulnerabilities they have).

The script in Listing 9-8 goes with option 1: embedding vulnerability identifiers in each host document. Host documents are likely to be longer-lived than vulnerability documents. If your hosts are subject to a regular patching regimen (which I understand is a tall order in some organizations), the host documents will remain for the long term. But the vulnerability documents, because they're patched on a regular basis, will age out of the database, using scripts you'll see in Chapter 10. If this assumption doesn't hold in your organization, you might want to use option 2.

MAPPINGS IN SQL

There is a third mapping option that makes sense when you're using SQL. If you store the mappings in a third table with 1:1 mappings of hosts and vulnerabilities, you only need to search in one place for both kinds of reporting. For example, "host A has vulnerability X" would be one record. Another would be "host A has vulnerability Y" and so on. When reporting, you'd first find the mappings you're interested in ("all vulnerabilities on host A") and then use the other two tables to flesh out the details of host A and its vulnerabilities X, Y, and Z. In a schematic, it would look something like this:

1. Query mapping table: find all records pertaining to host A and gather associated vulnerabilities in set B.
2. Query host table: find details of host A.
3. Query vulnerabilities table: find all records pertaining to vulnerability set B.

Experienced SQL users do this with JOIN statements in a single query; amateurs like me find it easier to run a few queries in sequence. This third solution is an example of *database normalization*. For more information, consult Wikipedia or your nearest computer bookshelf.

Building the Script

The script in Listing 9-8 iterates over the result tags, pulls the relevant data, and then sends that data to the database, keying off the OID as an authoritative vulnerability identifier, as discussed earlier in the section "Understanding the Database Structure" on page 81.

To build the vulnerability mapping, you must parse through the entire set of returned documents and build a list of which vulnerabilities apply to which hosts. Then replace the previous vulnerability list for each host with the new list.

```
#!/usr/bin/env python3

from xml.etree.cElementTree import iterparse
from pymongo import MongoClient
```

```
      import datetime, sys

      client = MongoClient('mongodb://localhost:27017')
      db = client['vulnmgt']

      # host - OIDs map
❶ oidList = {}

      def usage():
          print ('''
  Usage: $ openvas-insert.py <infile>
          ''')

      def main():
          if (len(sys.argv) < 2): # no files
              usage()
              exit(0)

          infile = open(sys.argv[1], 'r')

          for event, elem in iterparse(infile):

              if elem.tag == "result":
                  result = {}

              ❷ ipaddr = elem.find("host").text
                  (port, proto) = elem.find("port").text.split('/')
                  result['port'] = port
                  result['proto'] = proto
                  nvtblock = elem.find("nvt") # a bunch of stuff is in here

              ❸ oid = nvtblock.get("oid")
                  result['oid'] = oid
                  result['name'] = nvtblock.find("name").text
                  result['family'] = nvtblock.find("family").text

              ❹ cvss = float(nvtblock.find("cvss_base").text)
                  if (cvss == 0):
                      continue
                  result['cvss'] = cvss

                  # these fields might contain one or more comma-separated values.
                  result['cve'] = nvtblock.find("cve").text.split(", ")
                  result['bid'] = nvtblock.find("bid").text.split(", ")
                  result['xref'] = nvtblock.find("xref").text.split(", ")

              ❺ tags = nvtblock.find("tags").text.split("|")
                  for item in tags:
                      (tagname, tagvalue) = item.split("=", 1)
                      result[tagname] = tagvalue
                  result['threat'] = elem.find("threat").text
                  result['updated'] = datetime.datetime.utcnow()
                  elem.clear()
```

```
        ❻ if db.vulnerabilities.count({'oid': oid}) == 0:
                db.vulnerabilities.insert(result)


        ❼ if ipaddr not in oidList.keys():
                oidList[ipaddr] = []
            oidList[ipaddr].append({'proto': proto, 'port': port, 'oid': oid})


    ❽ for ipaddress in oidList.keys():
            if db.hosts.count({'ip': ipaddress}) == 0:
                db.hosts.insert({'ip': ipaddress,
                                    'mac': { 'addr': "", 'vendor': "Unknown"
},
                                    'ports': [],
                                    'hostnames': [],
                                    'os': [],
                                    'updated': datetime.datetime.utcnow(),
                                    'oids': oidList[ipaddress]})
            else:
                db.hosts.update_one({'ip': ipaddress},
                                    {'$set': {  'updated':
                                    datetime.datetime.utcnow(),
                                        'oids': oidList[ipaddress]}})

    infile.close() # we're done

main()
```

Listing 9-8: The openvas-insert.py *code listing for OpenVAS database insertion*

As with *nmap-insert.py* (Listing 9-5), you iterate over each result, collecting the information you require. First, you get the vulnerable host's IP
address ❷. Next, from subtags of the nvt tag, you get the OID (to identify
the vulnerability) ❸; the CVSS score (ignoring any vulnerabilities with a
CVSS score of 0) ❹; and the cve, bid, and xref fields (which contain one or
more comma-separated values). Then you get the key-value pairs from tags,
a free-form section in each vulnerability record that separates keys and val-
ues using a pipe character (|). Because you can't know ahead of time what
will or won't be in that field, this script simply parses all key-value pairs ❺
and adds them as is into the Mongo vulnerabilities document along with
the other data ❻. If the vulnerability already exists in the vulnerabilities
database, the script doesn't insert anything.

Then you add or update an entry in the host-to-vulnerability map
oidList ❶ for the host with information on each vulnerability found on that
host ❼. Once you're finished going through all the vulnerabilities, you can
use that map to add OIDs to each affected host document in hosts ❽ by
looping through the dictionary you created previously.

Customize It

If you find other information in the OpenVAS scan results useful, store that too. You could even store the entirety of the scan report data in a Mongo document. But you'd probably want to parse out items like the tags section into separate sections first. If you choose to go this route, it will take up a lot more space!

Because there's a lot of overlap in OpenVAS with the Nmap results, I skipped importing any results (like open ports) that would duplicate Nmap. You might want to supplement (or overwrite) the Nmap results or only use OpenVAS.

If you're interested in searching for specific vulnerability categories, you can expand the cvss_base_vector tag before creating the Mongo document (for example, "Access vector": "remote", "confidentiality impact": "high" and so on) by parsing this field in a similar way to the tags field, separating keys and values by the : and / characters.

Listing 9-8 uses OIDs as a unique identifier rather than BID/CVE IDs because not all scan results have the latter IDs but every result has an OID. Unfortunately, this leads to another problem: OpenVAS uses the same OID to track multiple instances of the same test on a host. For instance, it might run the "service detection" test on every open port on a host and report them all with the same OID but different descriptions. By only storing one document per OID, this script will overwrite these colliding reports. But it appears that this only occurs in low-importance (0.0 severity) tests, so I ignored those results entirely. Doing so might not suit your use case. I also considered (but eventually decided against) creating a hash from certain values—say, OID, summary, port, and description—and using that as the unique identifier. That way the script would only store a single instance of any given test result but not lose any data. If these low-severity test results are important in your environment, consider replacing OID with hash as the authoritative vulnerability identifier.

Similarly, my solution to the issue of stale vulnerability mappings— deleting all old mappings per host and replacing them—might not work in your environment, especially if you're pulling vulnerability data from multiple scanners. If you add a scanner tag to the vulnerability mappings in your host documents, you can delete only the appropriate mappings when importing new scan results.

Summary

In this chapter, you took your first steps toward building a working vulnerability management system. Congratulations! With the Nmap and OpenVAS data coming into your database, you can start generating simple reports that provide insight into your organization's current vulnerability status.

But before creating reports, you need to pause and do a few maintenance tasks. In the next chapter, you'll explore ways to improve your database structure and search time with indexes. You'll also write a script that automatically ages out old data to ensure that your reporting contains only fresh, actionable vulnerability information.

10

MAINTAINING THE DATABASE

Your database now contains information about hosts and vulnerabilities that you parsed out of Nmap and OpenVAS scan results. To generate meaningful reports, you need to start with a quality database that contains accurate, consistent, and recent information about your environment. Being able to generate those reports quickly is also beneficial. The examples in this chapter show you how to improve your database. You'll add indexes to speed query performance and maintain data integrity by restricting the values that you can place in the index. In addition, you'll automatically remove stale data so your reports are based only on the most recent vulnerability findings.

Defining Database Indexes

In Mongo, marking a key as an index indicates to the underlying system that you'll likely run many searches of that specific key-value pair across documents in the collection. For this reason, Mongo maintains an index of the values for that particular key to enable much faster search and retrieval.

You should add indexes to your document collections in Mongo for two reasons. First, by setting a certain key as an index, you can search against that key more quickly and efficiently. This practice will become more important as your database grows and you create analysis scripts that query the database heavily.

Second, indexes help with data integrity. An index with the unique property tells Mongo there can only be one document with a given key value, which makes the database resilient to accidental data duplication. For example, if you set the IP address key as a unique index, any attempts to add new documents with the same IP address as an existing document will result in an error.

Setting Indexes

You can set indexes using the createIndex command. For now, just set indexes for the uniquely identifying fields in each collection: IP address (for hosts) and OID (for vulnerabilities). The syntax to set an index is as follows:

```
db.hosts.createIndex({keyname:1}, {unique:1})
```

In this snippet, *keyname* indicates you're making an index, and unique:1 tells Mongo you want it to be a unique key.

Now open Mongo and set a couple of indexes, as shown in Listing 10-1.

```
$ mongo
> use vulnmgt
switched to db vulnmgt
> db.hosts.createIndex({ip:1}, {unique:1})
❶ {
    "createdCollectionAutomatically" : false,
    "numIndexesBefore" : 1,
    "numIndexesAfter" : 2,
    "ok" : 1
}
> db.vulnerabilities.createIndex({oid:1}, {unique:1})
{
    "createdCollectionAutomatically" : false,
    "numIndexesBefore" : 1,
    "numIndexesAfter" : 2,
    "ok" : 1
}
```

Listing 10-1: Creating indexes

Creating indexes can take a nontrivial amount of time depending on your document collection's size, which is all the more reason to create your indexes ahead of time! If the createIndex command is successful, it returns a JSON document with information about the index ❶.

Testing Indexes

After setting up your unique indexes, test them to ensure that you can't insert new documents using existing key values: find an existing key and then attempt to create a new document using that same key value, as shown in Listing 10-2.

```
> db.hosts.find({ip: "10.0.1.18"})
{ "_id" : ObjectId("57d734bf1d41c8b71daaee13"), "mac" : { "vendor" :
"Raspberry Pi Foundation", "addr" : "B8:27:EB:59:A8:E1" }, "ip" : "10.0.1.18",
"ports" : [ { "port" : "22", "state" : "open", "service" :
--snip--
> db.hosts.insert({ip:"10.0.1.18"})
WriteResult({
    "nInserted" : 0,
    "writeError" : {
        "code" : 11000,
      ❶ "errmsg" : "insertDocument :: caused by :: 11000 E11000 duplicate key
        error index: vulnmgt.hosts.$ip_1  dup key: { : \"10.0.1.18\" }"
    }
})
> db.vulnerabilities.find({oid:"1.3.6.1.4.1.25623.1.0.80091"})
{ "_id" : ObjectId("57fc2c891d41c805cf22111f"), "oid" :
"1.3.6.1.4.1.25623.1.0.80091", "summary" : "The remote host implements TCP
timestamps and therefore allows to compute\nthe uptime.", "cvss" : 2.6,
--snip--
> db.vulnerabilities.insert({oid:"1.3.6.1.4.1.25623.1.0.80091"})
WriteResult({
    "nInserted" : 0,
    "writeError" : {
        "code" : 11000,
      ❷ "errmsg" : "insertDocument :: caused by :: 11000 E11000 duplicate
        key error index: vulnmgt.vulnerabilities.$oid_1  dup key: { :
        \"1.3.6.1.4.1.25623.1.0.80091\" }"
    }
})
```

Listing 10-2: Testing uniqueness constraints

Mongo returns a duplicate key index error (❶, ❷) when you try to add new documents using the same unique index values as existing documents. This error test should protect you from any sloppy coding leading to duplication in your database collections.

Customize It

You can use other keys in your document structure as indexes. For instance, if you used a hash value to uniquely identify vulnerability results in Chapter 9, you would use that as an index rather than the OID.

If your database is already sizeable, or you expect it to become even larger before you determine the specific queries you'll be using in Chapters 11 and 13, you can read ahead now and determine which keys you want to index to save some time.

Keeping the Data Fresh

Your reports are only as good as the data they rely on, so it's important to make sure your information is up-to-date.

Your scripts to insert host information from Nmap and OpenVAS will update host information and add new hosts as needed. But what about old hosts? Imagine you scan a server in January and it appears in your database, but it gets decommissioned in February. You need to cull out-of-date information from your database to make sure you're reporting on actual hosts and vulnerabilities, not phantoms.

You can leave the vulnerabilities database as is because it contains vulnerability information that should be relatively static once added to the database. Vulnerability mappings are cleared and regenerated every time the vulnerability scan reports are imported (see "Getting OpenVAS into the Database" on page 91), so those shouldn't get stale as long as you regularly scan and ingest the results.

Determining the Cleanup Parameters

If you delete all entries that aren't found in the latest scan, you risk losing important data simply because a system was offline or unreachable at scan time. But if you keep data indefinitely, you might fill the database with irrelevant data, making it harder to find the active information you need. How long should you keep stale data? Or, how many scans can pass with no updates to a specific host before you decide it's gone and you can remove it?

The answer depends on how frequently you scan and any asset management policies in place at your organization. The script in the next section assumes that scans are weekly, and any information that hasn't been updated for one month (four scan cycles) is considered deprecated and removed.

The scripts so far insert all database entries with timestamps and update those timestamps whenever the data is updated. So you can use the timestamps to delete documents in the hosts database that haven't been updated in at least four weeks.

There are a few ways to perform this cleanup: you can run the commands manually from the Mongo command line, write a bash script that automatically runs the Mongo commands, or wrap the whole thing in a Python script. For the sake of consistency, we'll do the cleanup in Python.

Cleaning Up Your Database with Python

Listing 10-3 cleans up the database by deleting unchanged records older than a certain date. By default, this script assumes any data older than 28 days is stale and can be removed.

```python3
#!/usr/bin/env python3

# v0.2
# Andrew Magnusson

from pymongo import MongoClient
import datetime, sys

client = MongoClient('mongodb://localhost:27017')
db = client['vulnmgt']

❶ olderThan = 28

def main():

❷  date = datetime.datetime.utcnow()
❸  oldDate = date - datetime.timedelta(days=olderThan)


❹  hostsremoved = db.hosts.find({'updated': {'$lt': oldDate}}).count()
❺  db.hosts.remove({'updated': {'$lt': oldDate}})

❻  print("Stale hosts removed:", hostsremoved)

main()
```

Listing 10-3: A simple script (db-clean.py) that cleans up your database

This script calculates the current date ❷ and then uses olderThan ❶ to get the cutoff date for removing documents ❸. Then it queries the database to get a list of all documents where the value of updated is older than the cutoff date ❹, deletes those documents ❺ by telling Mongo to remove all, and prints the list of removed documents ❻.

Because this script assumes a 28-day deletion interval, you should run it at least once every 28 days by scheduling it with *cron*. We'll look at scheduling in more detail in Chapter 12.

Customize It

Customize the "stale data" variable (olderThan) in the script in Listing 10-3 to correspond to how often you run your Nmap and vulnerability scanner scans. We'll look at scan intervals in more depth in Chapter 12.

You can also use Mongo TTL indexes instead of a script. Adding a TTL index to your host collection tells Mongo to delete records automatically if, for instance, their updated field hasn't been changed in more than 28 days. But be careful not to let those deletion intervals happen while you're running analyses on your data.

Summary

Now that you've increased your database's speed and your data's quality, you can start looking at what your database can do for you. In Chapter 11, you'll learn how to extract information from your database and put it into human-readable form, starting with simple asset and vulnerability reports.

11

GENERATING ASSET AND VULNERABILITY REPORTS

Now that you have some asset and vulnerability data to work with, you'll use that data to generate reports about the asset information for each device in your database. When your boss asks, "How many Linux servers do we have?" or "How many of our desktops are susceptible to this new zero-day vulnerability I heard about on the news this morning?" you can use these reports to provide a confident answer. Before diving into this chapter, be sure to work through the database maintenance steps in Chapter 10.

Asset Reports

An *asset report* is an overview of all the different systems in your environment. It includes information about which OS and services each is running and how many vulnerabilities each has. Asset reports are useful when you want an overview of your organization's vulnerability environment on a host-by-host basis. Questions you can answer with this report include:

- How many hosts are in my environment?
- How many Linux servers do I have?
- How many vulnerabilities are on my production servers?
- Which of my workstations is most in need of patching?

Planning Your Report

When planning your reports, take the time to determine the information your report should contain. You already have an extensive amount of data, including a list of hosts with OS information, open ports, services, and vulnerabilities. You could just dump all of this data into an enormous spreadsheet, but then you'd have to sift through that data. Instead, we'll make a CSV file—a smaller and more readable spreadsheet—with only the most important data.

You can use Microsoft Excel or any other spreadsheet program to view and sort the data in a CSV file, create more detailed reports, or do further data analysis. For example, you might create a pivot table showing vulnerability counts per OS or summarize the asset list by OS or by open ports.

The script we'll use will collect the following information about each host in the database:

- IP address
- Hostname (if available)
- OS
- Open ports (TCP and UDP)
- Services detected
- Number of vulnerabilities found
- List of vulnerabilities by CVE

Table 11-1 shows examples of output for each column, with open ports, detected services, and the list of CVEs collapsed into lists of semicolon-separated values.

Collapsing fields with multiple values preserves the data in a searchable format at the cost of some readability. But it's a necessary compromise to represent all the data in a single CSV-formatted record.

Table 11-1: Asset Report Output Format

Column name	Example data
IP Address	10.0.1.1
Hostname	
OS	NetBSD 5.0 – 5.99.5
Open TCP ports	53; 5009; 10000;
Open UDP ports	
Detected services	domain; snet-sensor-mgmt; airport-admin;
Vulnerabilities found	1
List of CVEs	NOCVE

Getting the Data

Because we primarily want to focus on hosts, our first task will be to find a list of hosts. In this Mongo database, you're using IP addresses to uniquely identify each host, so you're guaranteed to have only one unique document per IP address.

To get a list of distinct IP addresses in the hosts collection, enter this command in the Mongo shell:

```
> db.hosts.distinct("ip")
```

Once you have that list, run the query in Listing 11-1, which uses find to get detailed information about a host with a given IP.

```
❶ > db.hosts.find({ip:"10.0.1.18"})
  {
    "_id": ObjectId("57d734bf1d41c8b71daaee13"),
    "mac": {
      "vendor": "Raspberry Pi Foundation",
      "addr": "B8:27:EB:59:A8:E1"
    },
    "ip": "10.0.1.18",
❷   "ports": [
      {
        "state": "open",
      "port": "22",
      "proto": "tcp",
      "service": "ssh"
      },
      {
        "state": "open",
        "port": "80",
        "proto": "tcp",
```

```
        "service": "http"
      },
  --snip--
      "updated": ISODate("2020-01-05T02:19:11.974Z"),
❸ "hostnames": [],
      "os": [
        {
          "cpe": [
            "cpe:/o:linux:linux_kernel:3",
            "cpe:/o:linux:linux_kernel:4"
          ],
    ❹ "osname": "Linux 3.2 - 4.0",
          "accuracy": "100"
        }
      ],
      "oids": [
        {
          "proto": "tcp",
    ❺ "oid": "1.3.6.1.4.1.25623.1.0.80091",
          "port": "general"
        }
      ]
```

Listing 11-1: Mongo output for one ip, reformatted for clarity

The output of the db.hosts.find query ❶ gives you open ports ❷ with port numbers, protocols proto, and service names; a list of detected hostnames (if any) ❸; the detected OS name (if any) ❹; and the OIDs of any vulnerabilities the scanner found on this host ❺.

You can look up each oid from the host document to get the associated CVE or CVEs using the script in Listing 11-2.

```
> db.vulnerabilities.find({oid:"1.3.6.1.4.1.25623.1.0.80091"})
{
  "_id": ObjectId("57fc2c891d41c805cf22111f"),
  "oid": "1.3.6.1.4.1.25623.1.0.80091",
  "summary": "The remote host implements TCP timestamps and therefore allows
  to compute\nthe uptime.",
  "cvss": 2.6,
  "vuldetect": "Special IP packets are forged and sent with a little delay in
  between to the\ntarget IP. The responses are searched for a timestamps. If
  found, the\ntimestamps are reported.",
  "bid": "NOBID",
  "affected": "TCP/IPv4 implementations that implement RFC1323.",
  "threat": "Low",
  "description": "It was detected that the host implements RFC1323.\n\nThe
  following timestamps were retrieved with a delay of 1 seconds in-between:\
  nPaket 1: 1\nPaket 2: 1",
  "proto": "tcp",
  "insight": "The remote host implements TCP timestamps, as defined by
  RFC1323.",
--snip--
```

```
    "impact": "A side effect of this feature is that the uptime of the remote\
    nhost can sometimes be computed.",
❶  "cve": [
      "NOCVE"
    ],
    "name": "TCP timestamps",
    "updated": ISODate("2020-10-11T00:04:25.601Z"),
    "cvss_base_vector": "AV:N/AC:H/Au:N/C:P/I:N/A:N"
}
```

Listing 11-2: Excerpted Mongo output for one oid, reformatted for clarity

Right now, we're only interested in the CVE or CVEs associated with this vulnerability ❶. But the output from Listing 11-2 provides a good deal of additional information that you can mine later. These general queries will obtain more information than you need, which is why we'll use the script to parse out the information we want to focus on.

There is one more wrinkle: as you saw in "Getting OpenVAS into the Database" on page 91, each OID might be associated with more than one CVE. Do you treat each associated CVE as a separate vulnerability in your vulnerability totals? Or, do you go by the OID count, which might be different from the count of specific CVEs? The script in the next section uses the OID count because it better captures the results the scanner returned. In general, multiple CVEs associated with a single OID are very similar. The alternative is to base the count on the total number of NOCVE or CVE-XXX-XXXX results discovered. This approach makes sense if you're only interested in unique results that are severe enough to have their own CVE identifier (not the numerous low-severity results OpenVAS will also return).

Script Listing

Listing 11-3 uses the queries we built in the previous section to generate the asset report.

```
#!/usr/bin/env python3
from pymongo import MongoClient
❶ import datetime, sys, csv
client = MongoClient('mongodb://localhost:27017')
db = client['vulnmgt']
outputFile = "asset-report.csv"
❷ header = ['IP Address', 'Hostname', 'OS', 'Open TCP Ports',
   'Open UDP Ports', 'Detected Services', 'Vulnerabilities Found',
   'List of CVEs']
def main():
    with open(outputFile, 'w') as csvfile:
        linewriter = csv.writer(csvfile)
    ❸  linewriter.writerow(header)
        iplist = db.hosts.distinct("ip")
    ❹  for ip in iplist:
            details = db.hosts.find_one({'ip':ip})
            openTCPPorts = ""
            openUDPPorts = ""
```

```
        detectedServices = ""
        serviceList = []
❺   for portService in details['ports']:
        if portService['proto'] == "tcp":
            openTCPPorts += portService['port'] + "; "
        elif portService['proto'] == "udp":
            openUDPPorts += portService['port'] + "; "
        serviceList.append(portService['service'])
❻   serviceList = set(serviceList)
    for service in serviceList:
        detectedServices += service + "; "
    cveList = ""
❼   if 'oids' in details.keys():
        vulnCount = len(details['oids'])
        for oidItem in details['oids']:
            oidCves = db.vulnerabilities.find_one({'oid':
            oidItem['oid']})['cve']
            for cve in oidCves:
                cveList += cve + "; "
    else:
        vulnCount = 0
❽   if details['os'] != []:
        os = details['os'][0]['osname']
    else:
        os = "Unknown"
    if details['hostnames'] != []:
        hostname = details['hostnames'][0]
    else:
        hostname = ""
❾   record = [ details['ip'], hostname, os, openTCPPorts,
        openUDPPorts, detectedServices, vulnCount, cveList]
    linewriter.writerow(record)
❿ csvfile.close()
main()
```

Listing 11-3: Script listing for asset-report.py

The script has two primary parts: the headers and declarations, and the main loop inside main(). To output to a CSV file, import the Python csv library ❶. The overall format of your CSV file is set with the header array ❷, which will be the first line to write to the output file ❸.

The main loop in the script ❹ goes through a list of unique IP addresses in the Mongo database and retrieves each IP's associated document. From the ports structure ❺, we collect all open ports (TCP and UDP) and service names, deduplicate the list of service names (because multiple ports might report the same service name) by casting it to a set ❻, and then write the ports and names to semicolon-separated strings. To get the vulnerability count and list, we check the host details for the oids key, count the OIDs found, and then query the vulnerabilities collection to get the corresponding CVE identifiers ❼. Next, we collect the OS name or, if none is included, report the detected OS as Unknown ❽. At the end, we collate all this information into a single line of the output CSV ❾ and write it to the output file. Then we move on to the next IP in the list. When the loop finishes with all the IP addresses, we close the CSV file, which writes the output to disk ❿.

Customize It

You can output the asset report as a nicely formatted HTML, PDF, or Word document. Modules for all three exist for Python. But reports in those formats aren't easily modified and sorted in another program the way that CSVs are. So you might sort the assets by IP, OS, or number of vulnerabilities in the script before writing the information to a file.

You can collect different details about your assets, depending on your needs, or manipulate certain fields further. For example, the detected OS names are often too granular (or not granular enough) to be useful for aggregation. So you can create a combined OS type field like "Windows" or "Linux" using string matching or regular expressions on the osname field to categorize each host under a more general OS type.

If you want to report on a subset of hosts, you can add logic either in the MongoDB query or in the Python script to select a subset of returned records. See Chapter 13 for more information on how to do this.

Vulnerability Reports

A *vulnerability report* is an overview of the specific vulnerabilities in your environment. This report is useful for addressing urgent vulnerabilities. For instance, if you can show that a particular vulnerability is widespread in your organization, you can build a case to do an emergency patch.

To generate the report, you'll write a script that looks for relevant data in your Mongo database and outputs it to a CSV file that is ready for further analysis in a spreadsheet program.

Planning Your Report

As with the previous script, you should first decide what you want to know about the vulnerabilities in your environment. Many fields in the vulnerability database (vulnmgt) that you imported from the scan results and in cvedb (this database is created by cve-search and is the source of most vulnerability details) aren't presently relevant. Which do we want to focus on? Items in the following list should be plenty for now:

- CVE ID
- Title (from cvedb)
- Description (from cvedb)
- CVSS score (from cvedb)
- CVSS details (from cvedb)
- Number of hosts with this vulnerability
- List of hosts with this vulnerability, by IP

The "number of hosts" field plus the CVSS field will let you prioritize vulnerabilities by sorting the results in a spreadsheet program.

Table 11-2 shows examples of output for each column, again with multiple values for one field collapsed as a semicolon-separated list.

Table 11-2: Vulnerability Report Output Format

Column name	Example data
CVE ID	NOCVE
Description	The remote host implements TCP time-stamps; allows for computing the uptime
CVSS	2.6
Confidentiality impact, Integrity impact, Availability impact	Partial, None, None
Access vector, Access complexity, Authentication required	Network, High, None
Hosts affected	1
List of hosts	10.1.1.31

Getting the Data

The vulnerability information is stored on a per-host basis and referenced by the OpenVAS OID. So you must first collect a list of OIDs from each host in your gathered scan results, cross-reference your vulnerabilities collection to determine the CVE (or lack thereof) for each vulnerability, and then get CVE details from the cvedb collection. In pseudocode, Listing 11-4 shows how the logic works.

```
For each host in hosts:
    Get all OIDs
    For each OID:
        Get CVE
        Associate CVE with host (CVE => (list of affected hosts) map)
For each CVE:
    Get cvedb fields
    Get and count hosts that are associated with this CVE
    Build output CSV line
```

Listing 11-4: Pseudocode for finding relevant CVEs

Building reverse correlations (CVE to host) on the fly avoids the overhead of a separate database collection that just has host and CVE pairs.

To handle the fact that some OIDs might have more than one CVE, you can separate OIDs that contain multiple CVEs into separate CSV lines. Or, you can choose only the first associated CVE, ignoring any remaining CVEs on the grounds that they're likely to be almost identical. The script in the following section separates OIDs into individual CVEs.

Script Listing

Listing 11-5 contains the code we'll use to generate a vulnerability report.

```python
#!/usr/bin/env python3
from pymongo import MongoClient
import datetime, sys, csv
client = MongoClient('mongodb://localhost:27017')
db = client['vulnmgt']
cvedb = client['cvedb']
outputFile = "vuln-report.csv"
header = ['CVE ID', 'Description', 'CVSS', 'Confidentiality Impact',
'Integrity Impact', 'Availability Impact', 'Access Vector',
'Access Complexity', 'Authentication Required', 'Hosts Affected',
'List of Hosts']
def main():
    with open(outputFile, 'w') as csvfile:
      linewriter = csv.writer(csvfile)
        linewriter.writerow(header)
          hostCveMap = {}
        hostList = db.hosts.find({'oids': {'$exists' : 'true'}})
❶ for host in hostList:
          ip = host['ip']
      ❷ for oidItem in host['oids']:
            cveList = db.vulnerabilities.find_one({'oid':
            oidItem['oid']})['cve']
              for cve in cveList:
                  if cve == "NOCVE":
                  continue
              ❸ if cve in hostCveMap.keys():
                    if ip not in hostCveMap[cve]:
                      hostCveMap[cve].append(ip)
                  else:
                    hostCveMap[cve] = [ ip ]
      ❹ for cve in hostCveMap.keys():
          cvedetails = cvedb.cves.find_one({'id': cve})
          affectedHosts = len(hostCveMap[cve])
        listOfHosts = ""
        for host in hostCveMap[cve]:
          listOfHosts += host + "; "
              if (cvedetails):
          if "impact" not in cvedetails:
              cvedetails["impact"] = {"availability": None,
              "confidentiality": None, "integrity": None }
          if "access" not in cvedetails:
              cvedetails["access"] = {"authentication": None,
              "complexity": None, "vector": None }
          record = [ cve, cvedetails['summary'], cvedetails['cvss'],
                  cvedetails['impact']['confidentiality'],
                  cvedetails['impact']['integrity'],
                  cvedetails['impact']['availability'],
                  cvedetails['access']['vector'],
                  cvedetails['access']['complexity'],
                  cvedetails['access']['authentication'],
                  affectedHosts, listOfHosts ]
```

```
        else:
            record = [ cve, "", "", "", "", "", "", "", "",
            affectedHosts, listOfHosts ]
        ❺ linewriter.writerow(record)
    csvfile.close()
main()
```

Listing 11-5: Script listing for vuln-report.py

Because this script's structure is so similar to *asset-report.py* (Listing 11-3), let's just look at the tricky parts. There are two main loops: one to go through each host and build the CVE-to-hosts map ❶ and the other to loop through the resulting map and output a line for each relevant CVE ❹.

The first loop goes through each host document to collect a list of OIDs ❷, resolving them to CVE IDs (if any) via the vulnerabilities collection. Then it builds a dictionary of CVE IDs mapped to a list of vulnerable hosts (identified by IP address). For each CVE, we check whether it's in the hostCveMap dictionary ❸. If it is, we then check whether the current IP is already mapped to the relevant CVE; if it's not, we add it to the list of IPs associated with that key. If the CVE isn't in our map and thus has no IPs associated with it, we create a new CVE key in the dictionary and create a list containing the current IP to use as the value. Once this map is complete, a second loop ❹ collects details from the cvedb database for each CVE key in hostCveMap. The vulnerability details, including the list of affected hosts collapsed into a single semicolon-separated list, are then written to a single CSV output line ❺.

Customize It

The method of vulnerability analysis I used in Listing 11-5 requires that you load the entire collection of hosts with vulnerabilities into RAM, which might be infeasible for large datasets of thousands of hosts with multiple vulnerabilities per host. To speed up the script, you can prebuild a collection with a 1:1 relationship between hosts and vulnerabilities and then query it for all hosts affected by a given CVE. You can do this either with a dedicated script or by modifying the *openvas-insert.py* script (Listing 9-8) to build this collection while you're parsing the OpenVAS output file. This spares your computer from having to load the entire collection of hosts with vulnerabilities into RAM. But you'll need to add some additional code to your other scripts, delete stale data, and ensure that relevant indexes are created correctly ("Defining Database Indexes" on page 98). Because a separate document collection will provide this mapping, you'll need to update your other scripts that insert and delete data to make them aware of this mapping.

As mentioned before, this database uses only CVSSv2 scores, because neither OpenVAS nor cve-search provides CVSSv3 scores. If CVSSv3 scores are important in your environment, use other data sources to fill in that gap.

In this script, I ignore all OpenVAS results that report a CVE of NOCVE. Generally, they're low-severity issues. If you want to include these in the report, you'll have to pull most of the fields for these results from the OpenVAS data rather than from the CVE database.

Summary

In this chapter, you built your first reports using the data in your vulnerability database. Because the two most important aspects of this database are the hosts (assets) and the vulnerabilities your scans discovered, it's only natural to report according to these two parameters.

In Chapter 12, we'll take a short side trip to fully automate the vulnerability scanning program. Then, in Chapter 13, you'll learn how to produce more complex reports on the data your scans collect.

12

AUTOMATING SCANS AND REPORTING

You've now created scripts to scan your network, inserted the results into your database, and generated simple reports from the resulting data. If you wanted to, you could run all those scripts manually every time you needed fresh information about your organization's vulnerability posture. But why do that when you can write another script to do the work for you? In this chapter, we'll automate this process using a bash script called *automation.sh*.

Automation might sound complicated, but a simple automation script just executes other scripts, one after the other, as in Listing 12-1, and is scheduled to run on a specific interval using *cron*.

```
#!/bin/bash
run-script-1
run-script-2
--snip--
run-script-x
```

Listing 12-1: A simple automation script

In our case, *automation.sh* will run the scripts we built in Chapters 8 through 11.

Visualizing the Automation Process

Before we build *automation.sh*, let's walk through the process from beginning to end so you're clear on the tasks you want to automate and in what order.

Figure 12-1 highlights the steps of the vulnerability management life cycle (described in Chapter 1) that we'll automate, which are collect data and analyze data.

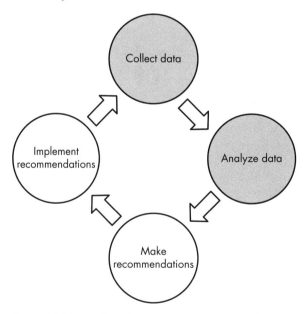

Figure 12-1: The vulnerability management life cycle

Collect Data

Collecting data usually occurs in two stages: run scans and then parse the results and import the data into your database. Chapters 8 and 9 showed the steps for this process.

1. Run Nmap; output the results to XML.

2. Run *nmap-insert.py* to parse the XML output and populate the Mongo database.

3. Run OpenVAS; output the results to XML.

4. Run *openvas-insert.py* to parse the XML output and populate the Mongo database.

Our automation script will use all of these steps in turn and will save the intermediate XML outputs with timestamps in case we need to review them later.

SERIAL OR PARALLEL OPERATION?

Instead of running one task, waiting for it to complete, and moving on to the next step, you could run some of the data-collecting steps in parallel. For example, you could run an Nmap scan and an OpenVAS scan at the same time, or you might import the results into Mongo in parallel. Monitoring multiple processes and ensuring they all finish before moving on to the next phase is an excellent challenge for experienced coders. But here we'll err on the side of a simple script that you can build and understand easily.

Analyze Data

By comparison, the *data analysis* step is easy: you run the scripts that generate reports (*asset-report.py* and *vuln-report.py* from Chapter 11) and then deliver the final result to whoever needs it. Of course, there's room for considerable complexity in this step, both within the reports that generate the results and in how you combine them. You might use other tools to generate more reports or perform your own analyses on the data. The types and number of reports you create will depend on the data you accumulate and the reason for the analysis.

Maintain the Database

Maintaining the database, which was discussed in Chapter 10, isn't part of the high-level vulnerability management process; nonetheless, it's important. This is a separate process you should do continually, so we'll build it into the automation script.

Planning the Script

The automation script doesn't need to be complex. All it has to do is run all the steps that until now you've been running manually and store the outputs for later investigation. The list of operations is straightforward: collect and then analyze that data—run your scans, import the data, run your reports, and handle database maintenance tasks along the way. But there are some nuances worth mentioning:

- **Order of operations.** You must collect data before you can insert it and insert data before you can generate reports. But you don't have to run your scans serially unless you customized your scan invocations so the input of one depends on the output of the other. For example, if you decided to save time in the OpenVAS script step by scanning only IP addresses that were returned as "live" by the Nmap scan, Nmap must finish before you start OpenVAS.

- **Short subscripts or direct command lines.** The database insertion process is complicated enough that it requires its own scripts. But since you can invoke Nmap using a single command, you can put the Nmap command directly into the main script. However, you might want to wrap that invocation in its own short script to keep your main script more consistent and readable. We'll consider this decision and how automating Nmap differs from automating OpenVAS in "Running Nmap and OpenVAS" later in this chapter.

- **Delivering the output.** A script that runs *asset-report.py* and *vuln-report.py* will generate the reports, but once you have them, you need to decide what to do with them. You might save them to a shared folder in your environment, upload them via a web form to a secure location, or just email them to yourself. Whatever you choose to do, make sure the reports end up in a place where you won't forget about them!

- **Keeping your environment organized.** Either delete the temporary files that scanning produces or save them in a way that won't be overwritten by newer scan results every time you run the script. If you maintain an organized environment, you can easily refer to old scan results when you need to look at them directly (instead of via the database).

- **Synchronizing with other scheduled tasks.** Don't schedule collection, reporting, updates, and maintenance tasks while you're in the middle of a system update (Chapter 7) or a database cleanup (Chapter 10). You should run these tasks when they won't conflict to avoid any incomplete or inaccurate results.

With these considerations in mind, here's a suggested outline for the *automation.sh* script. In the next section, I'll explain how I handled each design decision.

1. Run a database cleanup script to remove data older than one month (*db-clean.py*).
2. Run an Nmap scan on a configured network range; save the output to a timestamped XML file.
3. Import the results of the preceding Nmap scan into the database (*nmap-insert.py*).
4. Run an OpenVAS scan on a configured network range; save the output to a timestamped XML file.

5. Import the results of the preceding OpenVAS scan into the database (*openvas-insert.py*).

6. Run the reporting scripts; save the output to a timestamped CSV file.

Assembling the Script

Now that you know the steps your script needs to take and the order of the steps, you can build the script. Once it's assembled, you'll decide on an interval to run it, and you'll have a working vulnerability management system.

But before we get to the script listing, I want to highlight a few of the design decisions I made, keeping in mind the concerns I outlined in the previous section and the reasons for those decisions.

Running Nmap and OpenVAS

Nmap is very straightforward: all you need to do is run the tool from the command line with the required parameters. For that reason, I set up *automation.sh* to run Nmap directly and then run *nmap-insert.py* to store its XML output in the database.

OpenVAS is more complicated. You must start the scan with one omp command, then wait for it to finish (monitoring its process with another omp command) before running a third omp command to generate the XML output. It's possible to include all these steps directly in the automation script. But it's more modular, maintainable, and readable to break out the OpenVAS commands into their own script. So the *automation.sh* script needs to run the OpenVAS script (*run-openvas.sh*) and wait for it to complete before importing the resulting XML file into the database.

Scheduling the Script

To control precisely when the script runs, edit the system *crontab* directly. As root, add the following line at the bottom of */etc/crontab*, filling in the specific path to *automation.sh*:

```
4 0 * * 7 root </path/to/automation.sh>
```

This line will schedule the automation script to run at 12:04 AM, system time, on Sundays. You can also put the automation script, or a symlink to it, into */etc/cron.weekly* or whichever directory best suits your preferred scan interval.

Because you're also using *cron* to run your system update script from Listing 7-4, make sure the two scripts—system update and automation—don't run simultaneously. Most systems run entries in the */etc/cron.xxxx* directories in alphabetical order. But I'd suggest ensuring this is true in your environment before putting the scripts here for scheduling. If you're placing the scripts directly in *crontab*, leave a safe interval between the update and automation scripts, ideally running them on different days.

Script Listings

Listing 12-2, *automation.sh*, and Listing 12-3, *run-openvas.sh*, are the code listings you'll run. Don't forget to mark them as executable (run: chmod +x *filename*) before scheduling.

```
#!/bin/bash
❶ TS='date +%Y%m%d'
  SCRIPTS=/path/to/scripts
  OUTPUT=/path/to/output
  RANGE="10.0.0.0/24"
  LOG=/path/to/output-$TS.log
  date > ${LOG}
❷ echo "Running database cleanup script." >> $LOG
  $SCRIPTS/db-clean.py
❸ nmap -A -O -oX $OUTPUT/nmap-$TS.xml $RANGE >> $LOG
❹ $SCRIPTS/nmap-insert.py $OUTPUT/nmap-$TS.xml >> $LOG
❺ $SCRIPTS/run-openvas.sh >> $LOG
❻ $SCRIPTS/openvas-insert.py $OUTPUT/openvas-$TS.xml >> $LOG
  $SCRIPTS/asset-report.py >> $LOG
  mv $SCRIPTS/asset-report.csv $OUTPUT/asset-report-$TS.csv
  $SCRIPTS/vuln-report.py >> $LOG
  mv $SCRIPTS/vuln-report.csv $OUTPUT/vuln-report-$TS.csv
  echo "Finished." >> $LOG
```

Listing 12-2: Script listing for automation.sh

To timestamp the XML and CSV output files, we store the current time in a *YYYYMMDD* format in the variable TS ❶. Then we use the SCRIPTS and OUTPUT variables to store the paths to the script and output folders, respectively. We set RANGE to the network range or ranges we want to scan with Nmap. (Don't forget that OpenVAS is configured differently, via setting up targets, and is not bound by the range specified here.) We point LOG to a log file location that will also be tagged with the current timestamp. This file will hold the STDOUT output of each command for later review. Because every log and output file is timestamped, it will be easy to return to the script outputs later in case we run into problems or need to conduct additional analysis.

We run the database cleanup script ❷ to ensure that no stale data remains in the database. This script invokes Nmap directly ❸ but runs OpenVAS using the script shown in Listing 12-3 ❺. After running the database insert scripts ❹❻, we run the asset- and vulnerability-reporting scripts and move their output files to the OUTPUT directory. Then we add the line "Finished." to LOG so we know the script ran to completion.

Listing 12-3 presents the details of *run-openvas.sh*:

```
#!/bin/bash
❶ OUTPUT=/path/to/output
  TS='date +%Y%m%d'
❷ TASKID=taskid
  OMPCONFIG="-c /path/to/omp.config"
```

```
❸ REPORTID=' omp $OMPCONFIG --start-task $TASKID |
  xmllint --xpath '/start_task_response/report_id/text()' -'
❹ while true; do
      sleep 120
    ❺ STATUS='omp $OMPCONFIG -R $TASKID |
      xmllint --xpath 'get_tasks_response/task/status/text()' -'
      if [ $STATUS = "Done" ]; then
        ❻ omp $OMPCONFIG -X '<get_reports report_id="'$REPORTID'"/>'|
          xmllint --format - > $OUTPUT/openvas-$TS.xml
          break
      fi
  done
```

Listing 12-3: Wrapper script to run OpenVAS scans

This script assumes we already created a task, either through the Greenbone web GUI or the command line (see Chapter 8). By reusing the same task for each scan, we get a report history in the web GUI. We set the TASKID variable to the task *globally unique identifier (GUID)* (from the command line XML output or web GUI) ❷ and OMPCONFIG to the path of the configuration file with our OpenVAS credentials. Then we invoke omp to start the specified task ❸.

The omp command returns a chunk of XML that we'll parse with xmllint, the Swiss Army knife of XML tools. The --xpath flag tells it to return data from a specific location: the text content of a report_id tag inside a start_task_response tag. We save the resulting report ID to get the scan report later ❻.

The rest of the script is a simple loop ❹: wait two minutes; check the current status of the task ❺, again using xmllint; when it changes to Done, generate the final report ❻; and exit. At this point, a report will be in the configured output folder ❶, and the rest of the automation script will run as previously described in Listing 12-2.

Customize It

If you're using multiple scanners and have built your own scanning and database insertion scripts, you'll need to think about the order in which they execute and enter their results into the database. You'll need to consider these issues especially if you plan to have your scripts overwrite results that are already in the database.

If you want to run your Nmap and vulnerability scans separately or scan multiple network segments on different intervals, run them with separate scripts on their own intervals. Also, time your report generation so it doesn't run while a scan is still in progress or inserting its results into the database.

If you don't want to clear old results from your dataset every time you run new scans, schedule *db-clean.py* to run at a different interval instead of calling it from *automation.sh*.

Additionally, if you'd prefer not to synchronize your update script and your collection/reporting script(s), you can combine the two: first run a full update of your system, then run the scans and report on the results. Note that this will increase the script's total runtime.

Summary

In this chapter, you automated your scans and basic reporting, saving yourself a lot of busywork. At this point, you have a basic vulnerability management system: it periodically scans your environment and then generates and saves reports based on the most recent data available.

Now that you have a good understanding of the vulnerability management process, you can expand beyond the basics. In the next chapter, we'll look at more complex reports that you can generate from your scan data. Then, in Chapter 14, we'll incorporate other data sources and build a basic API to allow other tools to integrate with your vulnerability management system.

13

ADVANCED REPORTING

Now that you've completed the tedious aspects of assembling all the parts of your vulnerability management system—scanner, database ingester, and basic reporting—you can start building complex reports. So far, you've generated simple CSV tables of vulnerabilities and hosts. In this chapter, you'll expand your reports with more details about assets and vulnerabilities. You'll also integrate an additional data source (a list of exploitable vulnerabilities from the Exploit Database) to supplement your in-house data collection.

Detailed Asset Reporting

To expand on the basic asset report that contained system information, number of vulnerabilities, and CVE IDs of the vulnerabilities, you'll make three improvements:

- Add an option to report on a subset of hosts, selected by IP address.
- Enrich the report with all the host-specific data that's available, plus OpenVAS and cve-search data for each vulnerability associated with the host.
- Output the reports in HTML format because the expanded information won't map to a simple table.

To select hosts to include in the report, you'll add logic to your script to filter for a provided network range.

The following is the information you'll collect and format for each host:

IP address
The unique identifier for each host in the database.

Hostname(s)
In some environments, especially Windows-heavy networks where NetBIOS is in use, hostnames might be an easier way to identify individual hosts.

MAC information
Although this information is only worthwhile if the scanner can collect the host's MAC, this is another way to uniquely identify the host in question as well as learn about its networking hardware.

Detected OS (if multiple, choose the one with highest accuracy)
The underlying OS can help you sort your hosts as well as look at per-OS vulnerability trends.

Open ports (sorted by protocol, port number) and detected services
Just knowing which ports are open can be very useful to a security team.

Vulnerabilities associated with the host
All of the host's known vulnerabilities in one place, comprising the following details, all of which are pulled directly from the OpenVAS results:

- OID
- OpenVAS name
- OpenVAS summary
- OpenVAS CVSS score
- OpenVAS CVSS string
- CVE(s) associated

The resulting HTML document will look similar to Figure 13-1.

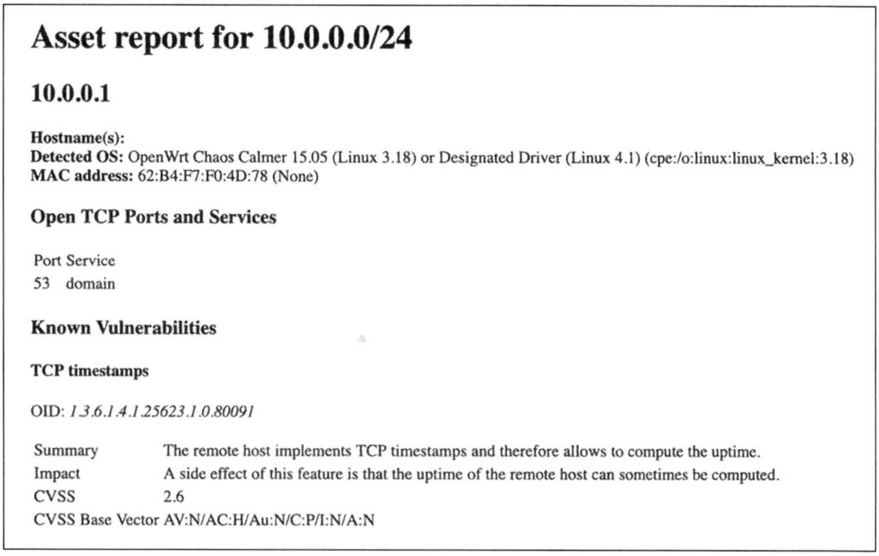

Asset report for 10.0.0.0/24

10.0.0.1

Hostname(s):
Detected OS: OpenWrt Chaos Calmer 15.05 (Linux 3.18) or Designated Driver (Linux 4.1) (cpe:/o:linux:linux_kernel:3.18)
MAC address: 62:B4:F7:F0:4D:78 (None)

Open TCP Ports and Services

Port Service
53   domain

Known Vulnerabilities

TCP timestamps

OID: *1.3.6.1.4.1.25623.1.0.80091*

Summary   The remote host implements TCP timestamps and therefore allows to compute the uptime.
Impact   A side effect of this feature is that the uptime of the remote host can sometimes be computed.
CVSS   2.6
CVSS Base Vector AV:N/AC:H/Au:N/C:P/I:N/A:N

Figure 13-1: Sample report for assets in range 10.0.0.0/24

Unlike the basic asset report, this one provides detailed information on a host-by-host basis that isn't structured in a spreadsheet format. Although the table borders are invisible, the "Open TCP Ports and Services" and "Known Vulnerabilities" sections are formatted via tables.

Planning the Script

Like your summary asset report (Listing 11-3 in Chapter 11), you'll draw largely from the hosts collection in Mongo in *detailed-assets.py*. But for each host, you'll also want information on every vulnerability (per OID) OpenVAS discovers for that host, which you'll pull from the vulnerabilities collection. Listing 13-1 shows the logic to put together all the information.

```
Get all unique hosts in 'hosts'
Filter per IP range, if any, passed as a command line parameter
For each host in 'hosts' after IP filter:
    Gather basic information in Mongo document
    For each OID associated with the host:
        Look up in 'vulnerabilities' collection
        Assemble data from returned document
    Format and present data
```

Listing 13-1: Pseudocode for detailed-assets.py

Once you've planned your script's logic, think about implementation details. How will you pass the filtering IP range to the script? Do you want to search for hosts any other way than by network range (more on this in "Customize It" on page 130)? How do you want to format the output HTML?

In Listing 13-2, I used the yattag library, which produces well-formatted HTML code by using standard Python structures and idioms. The yattag library automatically makes any text strings that pass through it HTML safe: for instance, it replaces < and > with < and >, and it replaces other entities with their HTML-encoded equivalents to ensure that the browser doesn't interpret any unexpected HTML code (or scripts!). If you manually generate the HTML tags, you must ensure that all strings pulled from the database are properly HTML formatted before placing them into the output file.

As we walk through the sample script that implements this logic, you'll learn about some of the other design choices I made. Use the code as inspiration to customize the form (displayed HTML) and function (specific fields retrieved from the database) for your own purposes.

Script Listing

Because the script *detailed-assets.py* in Listing 13-2 is fairly long, we'll walk through it in parts. First, we'll look at the script's preamble and table-setting portions: loading important libraries and starting the script with the main() function.

```
#!/usr/bin/env python3

from pymongo import MongoClient
from operator import itemgetter
❶ import datetime, sys, ipaddress
from yattag import Doc, indent

client = MongoClient('mongodb://localhost:27017')
db = client['vulnmgt']
outputFile = "detailed-asset-report.html"

def main():
❷   if len(sys.argv) > 1:
        network = sys.argv[1]
    else:
        network = '0.0.0.0/0'
    networkObj = ipaddress.ip_network(network)
❸   doc, tag, text, line = Doc().ttl()
❹   with tag('html'):
        with tag('head'):
            line('title', 'Asset report for ' + network)
```

Listing 13-2: Script listing for detailed-assets.py *(part 1)*

We import the ipaddress library (from Python 3) ❶ and the Doc and indent functions from yattag (installed via pip). If we pass the script no arguments, it reports on hosts in any network range. If we pass an argument, it's interpreted as an IP address range using *classless interdomain routing (CIDR)* notation ❷.

We initialize the yattag structure, which creates four objects (doc, tag, text, line) ❸ that will be used throughout the rest of the script to provide

HTML structures for the document, tags, text blocks, and short (one-line) tags, respectively. By using these four objects to generate chunks of HTML, the entire document is created in memory before it's written to a file at the end of the script. Keep this in mind because very large HTML documents for result sets of thousands or tens of thousands of assets might strain your vulnerability management system's available memory.

USING 'WITH'

The with structure lets you run code with setup and teardown procedures but abstracts away the complexity. When you run with function(): do things, you're actually running setupFunction(), do things, cleanupFunction().

The with tag('*tagname*') ❹ structure means the HTML tag *tagname* encapsulates all the output generated within the following indented code block. The with structure creates the tag (when it's invoked) and closes it when the block ends. In larger-scope tags, such as body and html, most of the script executes within the with block scopes!

Consider this brief example: imagine you want to make a simple HTML document, as shown in Listing 13-3.

```
<html>
    <head>
        <title>This is a title!</title>
    </head>
    <body>
        <h1>This is a heading!</h1>
        <p>This is some text in a paragraph.</p>
    </body>
</html>
```

Listing 13-3: Trivial HTML

You could use the yattag library to generate this using the Python snippet in Listing 13-4. Note that all you need to do is enclose the individual tag items (title, h1, p) within the larger with() blocks.

```
from yattag import Doc
doc, tag, text, line = Doc().ttl()
with tag('html'):
    with tag('head'):
        line('title', 'This is a title!')
    with tag('body'):
        line('h1', 'This is a heading!')
        line('p', 'This is some text in a paragraph.')
```

Listing 13-4: yattag example

Most of the rest of the script takes place within the context of the html tag. In the next code section (Listing 13-5) we retrieve the basic information for each host and place it in the in-progress HTML document.

```
❶ with tag('body'):
      line('h1', 'Asset report for ' + network)
      iplist = db.hosts.distinct("ip")
❷  iplist.sort(key=ipaddress.ip_address)
      for ip in iplist:
❸     if ipaddress.ip_address(ip) not in networkObj:
              continue
          details = db.hosts.find_one({'ip':ip})
          osList = details['os']
❹     if osList != []:
              osList.sort(key=itemgetter('accuracy'))
              os = osList[0]['osname']
              cpe = osList[0]['cpe'][0]
          else:
              os = "Unknown"
              cpe = "None"
          hostnameString = ""
❺     if details['hostnames'] != []:
              for name in details['hostnames']:
                  hostnameString += name + ', '

          line('h2', ip)
          line('b', 'Hostname(s): ')
          text(hostnameString)
          doc.stag('br')
          line('b', 'Detected OS: ')
          text(os + " (" + str(cpe) + ")")
          doc.stag('br')
          line('b', 'MAC address: ')
          if all (k in details['mac'] for k in ('addr', 'vendor')):
              text("{} ({})".format(details['mac']['addr'],
              details['mac']['vendor']))
          openTCPPorts = []
          openUDPPorts = []
❻     for portService in details['ports']:
              if portService['proto'] == "tcp":
                  openTCPPorts.append([int(portService['port']),
                  portService['service']])
              elif portService['proto'] == "udp":
                  openUDPPorts.append([int(portService['port']),
                  portService['service']])
          openTCPPorts.sort()
          openUDPPorts.sort()
          if len(openTCPPorts) > 0:
              line('h3', 'Open TCP Ports and Services')
              with tag('table'):
                  with tag('tr'):
                      line('td', 'Port')
                      line('td', 'Service')
                  for port, service in openTCPPorts:
```

```
                    with tag('tr'):
                        line('td', port)
                        line('td', service)

        if len(openUDPPorts) > 0:
            line('h3', 'Open UDP Ports and Services')
            with tag('table'):
                with tag('tr'):
                    line('td', 'Port')
                    line('td', 'Service')
                for port, service in openUDPPorts:
                    with tag('tr'):
                        line('td', port)
                        line('td', service)
```

Listing 13-5: Script listing for detailed-assets.py *(part 2)*

The with tag for the HTML body block is nested inside the html block ❶. We retrieve the list of hosts from the database and then sort it using the ipaddress library ❷. This is necessary because, unlike CSV, HTML can't be easily sorted. The ipaddress.ip_address is a sortable field, but if you're not using the ipaddress library, you'll need to write a custom sort function to account for the dot-decimal notation of IP addresses.

The main body of the script loops over IP addresses. We first check whether each host is in the provided IP range ❸. If it is, we generate a block of HTML with the host details from the database. OS detection can generate multiple options, so the script first sorts these by accuracy and then reports the first result ❹. (There might be multiple 100 percent accuracy guesses, so take this with a grain of salt!) Next, we create a list of all associated hostnames ❺, gather open ports, both TCP and UDP ❻, and then print a table of open ports for each protocol.

Listing 13-6 shows the last major output section: for each host, a list of vulnerabilities is associated with the host including basic details for each.

```
❶ if 'oids' in details:
        line('h3', 'Known Vulnerabilities')
        for oidItem in details['oids']:
            oidObj = db.vulnerabilities.find_one({'oid':
            oidItem['oid']})
            line('h4', oidObj['name'])
            with tag('p'):
                text('OID: ')
                line('i', oidObj['oid'])
            with tag('table'):
                with tag('tr'):
                    line('td', 'Summary')
                ❷ if 'summary' in oidObj:
                        line('td', oidObj['summary'])
                    else:
                        line('td', "")
                with tag('tr'):
                    line('td', 'Impact')
                    if 'impact' in oidObj:
```

```
                                line('td', oidObj['impact'])
                        else:
                                line('td', "")
                    with tag('tr'):
                        line('td', 'CVSS')
                        line('td', oidObj['cvss'])
                    with tag('tr'):
                        line('td', 'CVSS Base Vector')
                        line('td', oidObj['cvss_base_vector'])
            oidCves = db.vulnerabilities.find_one({'oid':
            oidItem['oid']})['cve']
            if oidCves != ['NOCVE']:
                line('h5', 'Associated CVE(s):')
                with tag('ul'):
                    for cve in oidCves:
                        line('li', cve)
        doc.stag('hr')
❸ with open(outputFile, 'w') as htmlOut:
        htmlOut.write(indent(doc.getvalue()))
        htmlOut.close()

main()
```

Listing 13-6: Script listing for detailed-assets.py *(part 3)*

We first check if any OIDs are associated with the host ❶ and then loop through them, gathering details. A number of the fields we want in our report are optional in the OpenVAS report, so you'll need to check whether any given tag exists before trying to insert it ❷. Once the HTML is fully generated, we write the complete document to the output file using another with tag ❸ and exit. The indent function makes the output more readable. The result is that we've generated the report shown in Figure 13-1. (If you take some time to refine the HTML document, you can even make it a little less ugly.)

Customize It

You could place the kind of structured data generated by *detailed-assets.py* into a Word document, a PDF, or a JSON structure to send to another system for further analysis. If you're a masochist, you could even structure it as a CSV table. But I'll leave representing this kind of nested data in CSV as an exercise for you.

Instead of filtering on IP range, you could filter on any of the fields in the host document, such as hostnames or os.cpe. You can add more filtering options to the command line or build the filters directly into the script if you're certain that you'll always want them for your detailed asset reports.

If you'd prefer not to install yet another Python library, you could hand generate your HTML tags by constructing strings and writing them to the output file. If you do so, you'll need to make all the data returned from Mongo, particularly the free-form text fields, HTML safe.

You can identify and sort hosts by hostname instead of IP address, especially in Windows-only environments where workstations and servers are often better known by hostname than IP address.

The script *detailed-assets.py* generates the entire HTML document in memory before writing it to a file at the end. If you have large datasets or limited RAM on your vulnerability management system, you can reduce RAM consumption by modifying the script to write the file piecemeal: first outputting the opening HTML tags and then one host record at a time.

In addition to using OpenVAS data, you can expand the script to pull data from the cve-search database for vulnerabilities with associated CVEs. You'll learn how to use this database in the next script.

Detailed Vulnerability Reporting

This script expands the simple vulnerability report from Listing 11-5 with detailed vulnerability information from OpenVAS and the cve-search database, compiled together in one readable HTML report. We'll add an IP range filter as a command line argument so you can view vulnerabilities for a subset of hosts. But the only host information in the report will be a list of IP addresses affected by each vulnerability and a count. We'll also add a filter to exclude vulnerabilities without CVE IDs: we'll assume that any vulnerability without a CVE isn't serious enough to describe in detail. If you want to see all the vulnerabilities in the database, just remove this filter.

Here is the information you'll collect and format for each vulnerability:

CVE
> The CVE ID of the vulnerability

Summary
> A brief description of the vulnerability

CWE
> The *common weakness enumeration (CWE)* category with a link to the online CWE database

Published date
> When the vulnerability was first disclosed publicly

Last update time
> The last time the vulnerability information was updated

CVSS score
> A numerical score from 0 to 10 of the overall vulnerability severity

CVSS details
> The individual breakdowns of each component comprising the full CVSS score, each given on a scale of "none, low, medium, high, critical":

- Confidentiality impact
- Integrity impact

- Availability impact
- Access vector
- Access complexity
- Authentication required

References
External links to reports, patches, and analysis

List of affected hosts
A list, by IP address, of hosts in your environment that this vulnerability affects

Count of affected hosts
The number of hosts with this vulnerability in your environment

As with CVE, the MITRE Corporation manages CWE and provides a comprehensive taxonomy of vulnerabilities. (If a vulnerability in some software you use is categorized, for example, as CWE-426: Untrusted Search Path, you can look up CWE-426 to learn how that class of vulnerabilities works.)

We include the list of references attached to the CVE, which might include patch or mitigation information, information about exploits, and third-party vulnerability reports, to provide important context that informs how your organization addresses the vulnerability.

The output will look similar to Figure 13-2. It's ordered by vulnerability, and the important details are output in an HTML table.

Vulnerability report for 10.0.0.0/24

CVE-2014-3120

Affected hosts: 1

Summary	The default configuration in Elasticsearch before 1.2 enables dynam MVEL expressions and Java code via the source parameter to _searc the user does not run Elasticsearch in its own independent virtual ma
CWE	CWE-284 (Improper Access Control)
Published	2014-07-28
Modified	2016-12-06
CVSS	6.8
Impacts	
Confidentiality	PARTIAL
Integrity	PARTIAL
Availability	PARTIAL
Access	
Vector	NETWORK
Complexity	MEDIUM
Authentication	NONE
References	
http://bouk.co/blog/elasticsearch-rce/	
http://www.exploit-db.com/exploits/33370	
http://www.rapid7.com/db/modules/exploit/multi/elasticsearch/script_mvel_rce	
http://www.securityfocus.com/bid/67731	

Figure 13-2: Sample output for detailed-vulns.py

Planning the Script

Because *detailed-vulns.py* is designed to return vulnerabilities found on hosts within a specific network range, the first step is to find all of those hosts. Once we have that list, we'll use it to find all vulnerabilities that exist on one or more hosts on the list. Then, using that list, we'll pull details for each vulnerability that has an associated CVE: the rest are ignored. Listing 13-7 shows what the logic looks like.

```
Get all unique hosts in 'hosts'
Filter per IP range, if any, passed as a command line parameter
For each host in 'hosts' after IP filter:
    Collect list of OIDs, insert into OID list
For each OID in OID list:
    Determine if it has a CVE; if not, go to next OID
    Gather data from associated CVE in cvedb database
    Format and present data
```

Listing 13-7: Pseudocode for detailed-vulns.py

Like *detailed-assets.py* in the preceding section, Listing 13-8 uses yattag to format and output the report in HTML. Because the structure is similar to the previous script, I'll provide the entire script in the next section and then draw your attention to a few key pieces.

Script Listing

Listing 13-8 shows the complete *detailed-vulns.py* script.

```
#!/usr/bin/env python3
from pymongo import MongoClient
import datetime, sys, ipaddress
from yattag import Doc, indent

client = MongoClient('mongodb://localhost:27017')
db = client['vulnmgt']
cvedb = client['cvedb']
outputFile = "detailed-vuln-report.html"

def main():
    if len(sys.argv) > 1:
        network = sys.argv[1]
    else:
        network = '0.0.0.0/0'
    networkObj = ipaddress.ip_network(network)
    hostCveMap = {}
    hostList = db.hosts.find({'oids': {'$exists' : 'true'}})
❶   for host in hostList:
        ip = host['ip']
        if ipaddress.ip_address(ip) not in networkObj:
            continue
        for oidItem in host['oids']:
            cveList = db.vulnerabilities.find_one({'oid':
            oidItem['oid']})['cve']
```

```
            for cve in cveList:
        ❷    if cve == "NOCVE":
                    continue
        ❸    if cve in hostCveMap.keys():
                    if ip not in hostCveMap[cve]:
                        hostCveMap[cve].append(ip)
             else:
                    hostCveMap[cve] = [ ip ]
    doc, tag, text, line = Doc().ttl()

    with tag('html'):
        with tag('head'):
            line('title', 'Vulnerability report for ' + network)
        with tag('body'):
            line('h1', 'Vulnerability report for ' + network)
        ❹ for cve in sorted(hostCveMap.keys()):
                cvedetails = cvedb.cves.find_one({'id': cve})
                affectedHosts = len(hostCveMap[cve])
                listOfHosts = hostCveMap[cve]
                line('h2', cve)
                line('b', 'Affected hosts: ')
                text(affectedHosts)
                doc.stag('br')
                if (cvedetails):
                    with tag('table'):
                        with tag('tr'):
                            line('td', 'Summary')
                            line('td', cvedetails['summary'])
                        with tag('tr'):
                            line('td', 'CWE')
                            with tag('td'):
                                id = 'Unknown'
                                if cvedetails['cwe'] != 'Unknown':
                                    id=cvedetails['cwe'].split('-')[1]
                            ❺ with tag('a',
                                href="https://cwe.mitre.org/data/"\
                                "definitions/"+id):
                                    text(cvedetails['cwe'])
                                cweDetails = cvedb.cwe.find_one({'id': id})
                                if cweDetails:
                                    text("(" + cweDetails['name'] + ")")
                                else:
                                    text("(no title)")
                        with tag('tr'):
                            line('td', 'Published')
                            line('td',
                            cvedetails['Published'].strftime("%Y-%m-%d"))
                        with tag('tr'):
                            line('td', 'Modified')
                            line('td',
                            cvedetails['Modified'].strftime("%Y-%m-%d"))
                        with tag('tr'):
                            line('td', 'CVSS')
                            line('td', cvedetails['cvss'] or 'Unknown')
```

```
                        with tag('tr'):
                            with tag('td'):
                                line('b', 'Impacts')
                        if 'impact' in cvedetails:
                            with tag('tr'):
                                line('td', "Confidentiality")
                                line('td', cvedetails['impact']
                                ['confidentiality'])
                            with tag('tr'):
                                line('td', "Integrity")
                                line('td', cvedetails['impact']['integrity'])
                            with tag('tr'):
                                line('td', "Availability")
                                line('td', cvedetails['impact']
                                ['availability'])
                        with tag('tr'):
                            with tag('td'):
                                line('b', 'Access')
                        if 'access' in cvedetails:
                            with tag('tr'):
                                line('td', "Vector")
                                line('td', cvedetails['access']['vector'])
                            with tag('tr'):
                                line('td', "Complexity")
                                line('td', cvedetails['access']['complexity'])
                            with tag('tr'):
                                line('td', "Authentication")
                                line('td', cvedetails['access']
                                ['authentication'])
                        with tag('tr'):
                            with tag('td'):
                                line('b', "References")
                        for reference in cvedetails['references']:
                            with tag('tr'):
                                with tag('td'):
                                    with tag('a', href=reference):
                                        text(reference)
                else:
                    line('i', "Details unknown -- update your CVE database")
                    doc.stag('br')

                line('b', "Affected hosts:")
                doc.stag('br')
                for host in sorted(listOfHosts):
                    text(host)
                    doc.stag('br')
    with open(outputFile, 'w') as htmlOut:
        htmlOut.write(indent(doc.getvalue()))
        htmlOut.close()

main()
```

Listing 13-8: Script listing for detailed-vulns.py

In the first main loop, we collect a list of vulnerabilities on the specified hosts by CVE ID ❶. If a vulnerability doesn't have an assigned CVE, we skip it ❷. We build a host map (a dictionary with CVE IDs as keys, mapped to a list of IP addresses) for each CVE during the first loop ❸, so when it comes time to list the hosts affected by each vulnerability, the information is already available. We loop through the full set of vulnerabilities ❹ and for each one generate a chunk of HTML containing vulnerability details (the same way as in *detailed-assets.py*). Because the vulnerability details include links to CWE information and CVE references, we need to use the HTML a tags with href attributes to generate links in the output report ❺.

Customize It

Several of the suggestions from *detailed-assets.py*, such as adding filters other than IP addresses and writing the report in chunks if it's large, might be useful to you in customizing this script as well.

You might want to include more host information, for example, hostname, total vulnerabilities on that host, and OS detection.

If you report on all discovered vulnerabilities, not just those with CVE IDs, you need data from the OpenVAS report to fill in for the data that's not in the cve-search database.

In addition, instead of sorting by CVE ID, from oldest to newest, you can sort from newest to oldest or use another sort entirely, such as total CVSS score.

Exploitable Vulnerability Reporting

Now that we've generated more complex reports, let's bring in external vulnerability information to enrich an existing report. In this example, you'll use a publicly available exploit repository, the Exploit Database (*https://exploit-db.com/*), and combine its information with the detailed vulnerability report to add another level of detail and actionability.

Preparation

To filter vulnerabilities according to their appearance in the Exploit Database, we'll use cve_searchsploit (available at *https://github.com /andreafioraldi/cve_searchsploit/*), a free command line tool, to search the Exploit Database. It contains a JSON file, *exploitdb_mapping_cve.json*, that directly maps CVE IDs to a list of exploits that apply to that CVE, which is exactly the data we need to add an exploitability filter to our vulnerability report.

To install cve_searchsploit, run this command:

```
$ git clone https://github.com/andreafioraldi/cve_searchsploit.git
```

This line installs the tool in the *cve_searchsploit/* subdirectory of the current directory. Once it's there, don't forget to add a command to your updater script (Listing 7-4) to periodically run `git fetch; git checkout origin/master -- cve_searchsploit/exploitdb_mapping_cve.json` within that directory to refresh the JSON file and ensure the mapping is up-to-date.

NOTE *The* References *section in the CVE database includes some Exploit Database links. But the mappings in* exploitdb_mapping_cve.json *are more comprehensive than the links in* References. *Using* cve_searchsploit *is a good example of integrating an external data source into our vulnerability management system.*

Modifying the Old Script

Because *exploitable-vulns.py* is essentially *detailed-vulns.py* with one more filter, which is to only report on vulnerabilities that also have known exploits in the Exploit Database, the changes required are minimal. We load the CVE-to-exploit map from *exploitdb_mapping_cve.json*, but before outputting a report on any given vulnerability, we ensure that it exists in that map. Then we add a section to the report with links to the exploits that also exist on the Exploit Database. Listing 13-9 shows the changes from *detailed-vulns.py*. You can find the complete script listing at *https://github.com/magnua /practicalvm/*.

```
#!/usr/bin/env python3
--snip other imports--
❶ import datetime, sys, ipaddress, json
--snip other global variables--
cveToExploitdbMap = "/home/andy/cve_searchsploit/cve_searchsploit/exploitdb_"\
"mapping_cve.json"

def main():
--snip network selection--
  ❷ with open(cveToExploitdbMap) as mapfile:
        exploitMap = json.load(mapfile)
--snip host finding--
    for host in hostList:
--snip CVE finding--
            for cve in cveList:
                if cve == "NOCVE":
                    continue
              ❸ if cve not in exploitMap:
                    continue
--snip CVE-to-host mapping--
    doc, tag, text, line = Doc().ttl()
    with tag('html'):
        with tag('head'):
            line('title', 'Exploitable vulnerability report for ' + network)
        with tag('body'):
            line('h1', 'Exploitable vulnerability report for ' + network)
            for cve in sorted(hostCveMap.keys()):
--snip most HTML generation--
                line('b', "ExploitDB links")
                doc.stag('br')
```

```
❹ for exploitID in exploitMap[cve]:
      with tag('a',
      href="https://www.exploit-db.com/exploits/"+exploitID):
          text("https://www.exploit-db.com/exploits/"+exploitID)
      doc.stag('br')
with open(outputFile, 'w') as htmlOut:
  htmlOut.write(indent(doc.getvalue()))
  htmlOut.close()

main()
```

Listing 13-9: Selected script listing for exploitable-vulns.py

First, we import the JSON library, which we need to parse `exploitdb_mapping_cve.json` ❶. Next, we load `exploitdb_mapping_cve.json` into memory, converting the JSON data into a Python dictionary using the `with` structure ❷. We discard vulnerabilities not in this map ❸ and convert the list of exploits into live links to their respective Exploit Database pages ❹.

Customize It

All of the suggestions from *detailed-vulns.py* are still valid for *exploitable-vulns.py*.

The Exploit Database is just one publicly available list of exploits. Another one is Metasploit, which I'll briefly discuss in the next chapter. You can import its vulnerability mappings from its local database by running the appropriate queries in *exploitable-vulns.py*.

You can also write filters similar to the exploitability filter for other vulnerability fields if you have matching external data sources. For example, you can import commercial vulnerability intelligence data to report on only vulnerabilities that are known to be under active attack by APT adversaries.

Summary

You now have several new and more complex reports to experiment with and customize. Don't forget to add them to your automation script (Listing 12-2) so they run regularly and you always have fresh reports.

At this point, your vulnerability management system is largely complete and, I hope, regularly generating useful vulnerability intelligence for your organization.

In Chapter 14, we'll look at system integration with your other tools via a basic application programming interface (API), automated exploitation and whether it makes sense in your environment, and vulnerability management systems in cloud environments.

14

ADVANCED TOPICS

You now have a fully functional and automated vulnerability management system. But projects like building this system are never actually finished. This chapter contains several ideas to enhance your system, including a simple integration API, automated penetration testing for known vulnerabilities, and cloud environments. Only the first script is a hands-on exercise: the rest discuss options and possibilities but leave the implementation details to you.

Building a Simple REST API

To get data from your vulnerability management system into another tool or to integrate your system into a third-party automation or orchestration product, you could do periodic database dumps, output reports in a format those tools can ingest, or write an API. If the destination tool supports API integration, using an API is a good solution. In this section, we'll look at building a simple representational state transfer (REST) API from scratch. But first, let's look at what a REST API is.

An Introduction to APIs and REST

Programmatic interfaces (shared boundaries between system components that are accessed using programs) provide a consistent method for programs to interact with each other and with the host OS. When you use an API, you don't need to understand the inner workings of the application you're communicating with; you just need to know that if your program writes *this* message to *that* location, the receiving system will understand and respond with a response of *that* type. Abstracting the inner workings behind an interface that remains consistent, no matter how the infrastructure behind that interface might change, greatly simplifies software development and interoperation. The reason is that programs can evolve independently while retaining a common communication language.

REST defines a class of APIs that communicate over the internet by reading from and writing to an unknown database (or other arbitrary storage system). A full REST API supports all database operations: creating, reading, updating, and deleting records (commonly called CRUD). The HTTP methods POST, GET, PUT (sometimes PATCH), and DELETE implement their respective CRUD operations, as shown in Table 14-1.

Table 14-1: HTTP Methods Mapped to CRUD Actions

Method	Action
GET	Get (read) the contents of a record (or information about multiple records)
POST	Create a new record
PUT/PATCH	Update an existing record or create one if it doesn't yet exist
DELETE	Delete an existing record

To use the API, the client sends an HTTP request using the appropriate method to a URL (technically, a *universal resource indicator (URI)*) that specifies the record or records to act on. For example, I send a GET request to *http://rest-server/names/* to tell the REST API to send back a list of names (commonly in XML or JSON). A GET request to *http://rest-server/names/andrew-magnusson/* returns more information about the name record for "Andrew Magnusson." A DELETE request to that same address tells the remote system to delete my name record.

The address in a URI, unlike one in a standard web URL, doesn't point to a consistent web location. Instead, it points to an *API endpoint*: an interface for a program running on the server side that the REST client uses to send the appropriate HTTP method to perform CRUD actions.

Designing the API Structure

Think about what you need your vulnerability management API to do. How many of the CRUD actions will you support? In *simple-api.py*, I implement only GET (read existing records), the simplest and safest method; all a client can do is request data that's already in the database. Our vulnerability management system updates itself internally, so there's no need for external systems to make changes to the database. If you want external systems (particularly automation or orchestration routines) to modify the vulnerability database, you can implement POST, PUT/PATCH, and DELETE methods.

You also need to consider what data the API clients should have access to. Your vulnerability management database contains a list of hosts with associated details, a list of discovered vulnerabilities with their own details, and a mirror of the CVE database provided by cve-search. We don't need to provide the CVE database contents with our API because it's publicly available. If other tools need this information, there are easier ways for them to get it than by querying your API. But it makes sense to expose host and vulnerability information that is specific to your network and most likely can't be found anywhere but the vulnerability management system.

The *simple-api.py* implements four endpoints for the hosts and vulnerabilities collections, accessible only via the GET method. Table 14-2 lists the details of each endpoint.

Table 14-2: API Endpoints and Their Function

Endpoint	Action
/hosts/	Returns a JSON-formatted list of IP addresses in the database
/hosts/<ip address>	Returns JSON-formatted host details for the provided IP address, including a list of CVEs it's vulnerable to
/vulnerabilities/	Returns a JSON-formatted list of CVE IDs in the vulnerabilities database; that is, CVEs that currently affect hosts in the system
/vulnerabilities/<CVE ID>	Returns JSON-formatted details for the provided CVE ID, including a list of IP addresses that are vulnerable

If any other URI paths are requested from the server where you host your API, the script returns a JSON document containing a key-value pair in the form {'error': 'error message'} and an HTTP status code. An HTTP status code of 2*xx* indicates success, and the 4*xx* series refers to a variety of errors (for example, 404 "page not found"). For purely whimsical reasons, I decided to make all the errors return code 418, which in HTTP unofficially means (I'm not making this up) "I'm a teapot." Feel free to use a different error code in your script.

Implementing the API

Instead of building the entire API in a single main() function, we'll split the script into logical functions:

main()   Starts the server instance and tells Python to handle all requests via SimpleRequestHandler.

SimpleRequestHandler   A custom class that inherits from the http.server .BaseHTTPRequestHandler class and overrides the do_GET function that parses the request URI for GET requests. It either returns an error or passes control to the database lookup functions that handle requesting and parsing data from Mongo. It returns an error for other HTTP method handlers like do_POST and do_PUT because they're not supported.

Database lookup functions   There are four of these, one for each endpoint. Each one performs Mongo queries and returns the data to SimpleRequestHandler in a JSON document, as well as a response code in the case of errors.

We'll look at each section in order, starting with the Python headers and the main() function in Listing 14-1.

```python
#!/usr/bin/env python3

❶ import http.server, socketserver, json, re, ipaddress
from bson.json_util import dumps
from pymongo import MongoClient
from io import BytesIO

client = MongoClient('mongodb://localhost:27017')
db = client['vulnmgt']
❷ PORT=8000
ERRORCODE=418 # I'm a teapot

--functions and object definitions are in Listings 14-2 and 14-3--

❸ def main():
    Handler = SimpleRequestHandler
    with socketserver.TCPServer(("", PORT), Handler) as httpd:
        httpd.serve_forever()

main()
```

Listing 14-1: Script listing for simple-api.py *(part 1)*

We import http.server and socketserver for basic HTTP server functionality, bson.json_util for a BSON-dumping utility to turn Mongo responses into clean JSON, and BytesIO to build the server response, which must be in a byte format rather than simple ASCII ❶. The global variables PORT and ERRORCODE ❷ define the listening port for the server and the standard error code to return, respectively.

When the script starts ❸, we instantiate a TCPServer, listening at the configured port. It delegates its handling to SimpleRequestHandler and, because it's invoked with serve_forever, will continue serving requests until the process is killed.

When a request comes in via GET, the do_GET method of SimpleRequestHandler in Listing 14-2 kicks into action.

```
class SimpleRequestHandler(http.server.BaseHTTPRequestHandler):
    def do_GET(self):
    ❶ response = BytesIO()
    ❷ splitPath = self.path.split('/')
        if (splitPath[1] == 'vulnerabilities'):
            if(len(splitPath) == 2 or (len(splitPath) == 3  and splitPath[2]
            == '')):
                self.send_response(200)
              ❸ response.write(listVulns().encode())
            elif(len(splitPath) == 3):
              ❹ code, details = getVulnDetails(splitPath[2])
                self.send_response(code)
                response.write(details.encode())
            else:
              ❺ self.send_response(ERRORCODE)
                response.write(json.dumps([{'error': 'did you mean '\
                'vulnerabilities/?'}]).encode())
        elif (splitPath[1] == 'hosts'):
            if(len(splitPath) == 2 or (len(splitPath) == 3  and splitPath[2]
            == '')):
                        self.send_response(200)
              ❻ response.write(listHosts().encode())
            elif(len(splitPath) == 3):
              ❼ code, details = getHostDetails(splitPath[2])
                self.send_response(code)
                response.write(details.encode())
            else:
                self.send_response(ERRORCODE)
                response.write(json.dumps([{'error': 'did you mean '\
                'hosts/?'}]).encode())
        else:
            self.send_response(ERRORCODE)
            response.write(json.dumps([{'error': 'unrecognized path '
            + self.path}]).encode())
        self.end_headers()
    ❽ self.wfile.write(response.getvalue())
```

Listing 14-2: Script listing for simple-api.py *(part 2)*

To determine whether the request path is one of the four supported endpoints, the requested URI is first split into its component pieces. To handle this parsing, we split the path into an array using / (forward slash) as the delimiter ❷. The first value in that array is blank (the empty string prior to the first slash), so the second and third values point to the appropriate database lookup function ❸❹❻❼, and the return values of those functions are used as the response body. If no function is matched, an error is returned as the response. Building the response in http.server requires three steps (four if any errors are generated):

1. Send headers (implicitly handled by sending a response code with send_response).
2. End headers with end_headers().
3. Generate errors as needed using ERRORCODE ❺.
4. Send actual response data with wfile.write ❽, which takes the byte-stream from the response variable. This variable is instantiated as a BytesIO object ❶ and is built by adding data to it via response.write, which automatically puts it into the proper byte format.

Additionally, there are four database functions: listHosts, listVulns, getHostDetails, and getVulnDetails, as shown in Listing 14-3.

```
def listHosts():
 ❶ results = db.hosts.distinct('ip')
    count = len(results)
    response = [{'count': count, 'iplist': results}]
 ❷ return json.dumps(response)

def listVulns():
    results = db.vulnerabilities.distinct('cve')
    if 'NOCVE' in results:
        results.remove('NOCVE') # we don't care about these
    count = len(results)
    response = [{'count': count, 'cvelist': results}]
    return json.dumps(response)

def getHostDetails(hostid):
    code = 200
    try:
     ❸ ipaddress.ip_address(hostid)
     ❹ response = db.hosts.find_one({'ip': hostid})
        if response:
            cveList = []
         ❺ oids = db.hosts.distinct('oids.oid', {'ip': hostid})
            for oid in oids:
                oidInfo = db.vulnerabilities.find_one({'oid': oid})
                if 'cve' in oidInfo.keys():
                    cveList += oidInfo['cve']
            cveList = sorted(set(cveList)) # sort, remove dupes
            if 'NOCVE' in cveList:
                cveList.remove('NOCVE') # remove NOCVE
         ❻ response['cves'] = cveList
        else:
            response = [{'error': 'IP ' + hostid + ' not found'}]
            code = ERRORCODE
    except ValueError as e:
        response= [{'error': str(e)}]
```

```
        code = ERRORCODE
    return code, dumps(response)

def getVulnDetails(cveid):
    code = 200
❼  if (re.fullmatch('CVE-\d{4}-\d{4,}', cveid)):
❽      response = db.vulnerabilities.find_one({'cve': cveid})
        if response: # there's a cve in there
            oid = response['oid']
❾          result = db.hosts.distinct('ip', {'oids.oid': oid})
            response['affectedhosts'] = result
        else:
            response = [{'error': 'no hosts affected by ' + cveid}]
            code = ERRORCODE
    else:
        response = [{'error': cveid + ' is not a valid CVE ID'}]
        code = ERRORCODE
    return code, dumps(response)
```

Listing 14-3: Script listing for simple-api.py *(part 3)*

The first two database functions query Mongo for a complete and dedu-plicated list of IP addresses (listHosts) ❶ or CVE IDs (listVulns) and send it back as a JSON structure ❷.

The *details* functions first validate whether the input value is a legiti-mate IP address ❸ or CVE ID ❼ and send back an error if not. Next, they pull out the details for a specific host ❹ or vulnerability ❽. Then they run a second query to get the list of associated hosts (for a vulnerability) ❾ or vul-nerabilities (for a host) ❺. This data, once collected, is inserted into a JSON structure ❻ that is returned to SimpleRequestHandler and then the client.

Getting the API Running

Once the *simple-api.py* script is complete and tested, set it up on your server to run all the time. The process for doing this depends on the service man-agement system that your OS uses: common ones for Linux are *systemd*, *SysV-style init*, and *upstart*. These instructions apply to *systemd*.

Create a service file called simple-api.service in the *systemd* scripts location (/lib/systemd/system on Ubuntu) to add a new *systemd* service. Listing 14-4 shows the contents of the service file.

```
[Unit]
Description=systemd script for simple-api.py
DefaultDependencies=no
Wants=network-pre.target

[Service]
Type=simple
RemainAfterExit=false
ExecStart=/path/to/scripts/simple-api.py
ExecStop=/usr/bin/killall simple-api
```

```
TimeoutStopSec=30s

[Install]
WantedBy=multi-user.target
```

Listing 14-4: Service configuration for simple-api.py

Now make *simple-api.py* executable using chmod +x and run the commands in Listing 14-5 as root to start the service and ensure that it's running.

```
# systemctl enable simple-api.service
Created symlink /etc/systemd/system/multi-user.target.wants/simple-api.service
→ /lib/systemd/system/simple-api.service.
# systemctl daemon-reload
# systemctl start simple-api
# systemctl status simple-api
  simple-api.service - SystemD script for simple-api.py
   Loaded: loaded (/lib/systemd/system/simple-api.service; enabled; vendor
   preset: enabled)
   Active: active (running) since Sun 2020-04-26 16:54:07 UTC; 1s ago
 Main PID: 1554 (python3)
    Tasks: 3 (limit: 4633)
   CGroup: /system.slice/simple-api.service
           1554 python3 /path/to/scripts/simple-api.py

Apr 28 16:54:07 practicalvm systemd[1]: Started systemd script for
simple-api.py.
```

Listing 14-5: Starting the service

First, systemctl enable adds *simple-api.service* into the *systemd* configuration. Next, systemctl daemon-reload and systemctl start simple-api start the service. Then systemctl status simple-api outputs the response you see in Listing 14-5 if the service is successfully running. At this point, the API will be up and listening on the port you've configured within the script.

Customize It

Python's *http.server* library minimizes external dependencies and makes it very clear how the code functions. But it doesn't provide API-specific functionality and only supports basic HTTP authentication (the Python authors strongly recommend that you not use it in a production environment). If you want to significantly expand the API, you can use a REST framework, such as Flask or Falcon, to simplify coding and maintain the API.

The *simple-api.py* script doesn't even implement basic HTTP authentication. So it's very important to either heavily restrict access to the web server or add authentication to the script before using it in production.

The script returns a simple list of hosts or vulnerability IDs from the */list* endpoints. You can return more information about every host/vulnerability, similar to the advanced reports in Chapter 13.

If you expect clients to use your API by requesting large batches of data, you can make this easier and more efficient by adding the option to include paging information in the query. For example, a request to *http://api-server/hosts/list/?start=20&count=20* would return records 20 through 40, and a client could iterate through the total host listing a batch at a time.

As the script and *systemd* service are written now, the log messages from `http.server` are printed to `STDERR`, which may not be captured by the *systemd* logger, *journald*. You can modify the script or the service definition to retain logs so you can keep an eye on who is using your API.

Descriptive error messages let an attacker probe your API to see what information is available. You can harden the API by replacing all the errors with a generic message that doesn't provide hints to the correct endpoint formats.

Automating Vulnerability Exploitation

Once you have information about systems containing vulnerabilities with known exploits (see Listing 13-9), you can determine whether those vulnerabilities are exploitable. If they are, you might prioritize fixing or mitigating those vulnerabilities. If they're not, either it's a false positive result or existing mitigations protect the host from successful exploitation.

But this process is slow and tedious: you have to find the exploit, set up your system to run it, attempt exploitation, and then record the results. You've already automated most of your process, so why not automate this final step as well? Tools like Metasploit are scriptable via the command line, so is there any reason not to automatically attempt exploitation?

Pros and Cons

Actually, there are several very good reasons not to automate vulnerability exploitation. Even the process of vulnerability scanning isn't without risks. It's always possible to cause glitches or even crash a system with aggressive scans or fragile targets. Running exploits is more dangerous yet: they could crash a production system, damage important data, or even (in rare cases) damage the underlying hardware. Exploit code that you don't thoroughly understand might have backdoor functions or unexpected side effects. Even if you could ensure that the exploits you're running do nothing but verify that exploitation is possible, you could still damage the system you're testing.

For many organizations, the risk isn't worth the reward of knowing which systems in the environment are vulnerable to which exploits. So they perform vulnerability exploitation manually, or at least partially manually. It's best to have an experienced penetration tester attempt exploitation in a controlled environment. The tester uses an exploitation framework like Metasploit to automate tedious steps, such as running tests repeatedly with different inputs or trying different exploits until they find one that works. But there's always a human monitoring its effectiveness and ready to stop the test if something goes wrong.

Some organizations have a large set of assets and a threat model where the exploitation risk is significantly higher than the risk of occasionally crashing a critical service. If manual exploitation of all the critical vulnerabilities isn't feasible, the additional information might be worth the risk. But this isn't a decision you should make lightly or in a vacuum. You'll need organizational support before implementing automatic vulnerability exploitation (see "Gaining Support" on page 39).

Automating Metasploit

Once you've identified which exploits exist for vulnerabilities in your environment, you need to run the identified exploits against the vulnerable host. With the Exploit Database, there's no easy way to script "run this exploit on host X": exploits are written in various languages, some might need to be compiled before running, and they have received varying levels of vetting for effectiveness and safety. As a unified penetration-testing framework, Metasploit solves these issues. All Metasploit-compatible exploits are implemented in Ruby, tested extensively, and run in a consistent manner via the Metasploit Framework. Better still, you can script Metasploit from the command line and encapsulate it in a Python (or similar) script. This section describes how to write such a script, but I'll leave the implementation as an exercise for the motivated reader.

NOTE   *You can modify the* exploitable-vulns.py *script in Listing 13-9 to use Metasploit's internal vulnerability-to-exploit mapping and be confident that any systems thereby marked as exploitable do in fact have automatable Metasploit modules. Access to this data and parsing it to find those mappings is another exercise I'll leave to the advanced reader.*

Listing 14-6 shows the overall structure of a possible automated exploitation script in pseudocode.

```
Query database for list of hosts with vulnerabilities
Map vulnerabilities against list of exploits (Exploit-DB, Metasploit, other)
Result: list of hosts and exploitable vulnerabilities on each host
For each host in this list:
    For each vulnerability on that host:
        Determine Metasploit module for specified vulnerability
        Kick off Metasploit module against specified host
        Record success/failure in host record in database
```

Listing 14-6: Pseudocode for automated exploitation with Metasploit

Getting a list of exploitable vulnerabilities on each host by mapping them against an existing list of exploits should be familiar from working with *exploitable-vulns.py*. The loop in Listing 14-7 goes through each exploitable vulnerability on each host and starts a Metasploit session to attempt to exploit the vulnerability with its associated Metasploit module.

Because Metasploit modules are referred to by name rather than by CVE ID, you'll need to connect the CVE you're attempting to exploit with the correct module. If you're not getting exploit information from Metasploit, you can correlate CVE IDs with Metasploit module names by manually parsing Metasploit searches, as in Listing 14-7.

```
$ msfconsole -qx 'search cve:CVE-2012-2019;quit'

Matching Modules
================

   #  Name  Disclosure Date  Rank  Check  Description
   -  ----  ---------------  ----  -----  -----------
   1  exploit/windows/misc/hp_operations_agent_coda_34  2012-07-09  normal
   Yes    HP Operations Agent Opcode coda.exe 0x34 Buffer Overflow
```

Listing 14-7: Searching Metasploit modules using the Metasploit command line

This process takes quite some time, mostly because starting `msfconsole` can take tens of seconds. You can split this listing into two scripts: one to start `msfconsole` and the other to submit requests to the running console process via a simple API.

Once you have the module name, the remaining step is to attempt exploitation. Run `msfconsole -qx 'command1;command2;commandX;quit'` to run a sequence of exploit-related commands and then close Metasploit. Many modules require additional parameters for the best operation: you might decide to run every module with its default configuration or store parameters for some of the more popular modules separately. To determine whether exploitation was successful, you can rely on the Metasploit output. Or, if you've configured Metasploit to use a database, you can pull success/failure information from the database after the exploit was attempted.

At this point, you can test automatic exploitation. But before you do so, consider the following:

- Is automatic exploitation testing necessary?
- Can I run this script against a test environment configured to replicate the live environment rather than against production systems?
- Is this testing *really* necessary?

If you're still convinced, good luck, and have at it!

Bringing the System into the Cloud

This book focuses on small organizations with on-premise workstations and servers. But businesses are increasingly adding cloud-based operations or even moving their entire production environment into the cloud. Many new organizations are forgoing local infrastructure entirely, opting to place their entire business infrastructure in a cloud environment. In this section, we'll look at some considerations for adding your cloud environment into your existing vulnerability management system.

Cloud Architecture

If your infrastructure is entirely in the cloud, it makes sense to deploy your vulnerability-scanning system entirely in the same cloud environment. Doing so will minimize latency and let you allow access to your various cloud network segments from a scanner that's already in the same environment.

But if your environment is a mix of cloud and on-premise infrastructure, you might need to consider a few different options. You could set up your cloud environment to permit your scanning tools access into the cloud. Or, you could set up separate scanners within the cloud environment that deliver their results to your centralized Mongo database. Scanning the cloud environment from a local scanner introduces latency (especially if you're geographically distant from your cloud network) and intervening security devices. You'll have to allow your scanner unlimited egress from your local network and permit its public IP address unlimited access to the cloud environment. Alternatively, you could provide this access via a virtual private network (VPN) configuration, which would let you securely tunnel traffic between your local and cloud environments.

If you set up multiple scanners for cloud or a heavily segmented local network, you'll need to ensure they coordinate their database insertions to avoid overwriting each other's data. You also must make sure that database reporting and deletion only happens from one location to guarantee the data remains consistent.

Cloud and Network Ranges

Unlike an on-premise network, where you know that all the IP addresses in a range are part of your network, cloud hosts or services often have multiple IP addresses: at a minimum, one for private access from within the same network and one for public internet access. In the private address space, cloud network separation ensures that you can't target hosts belonging to another cloud environment. But with public addresses, there is no such guarantee: your cloud's public IP addresses are adjacent to many other addresses.

If you scan only your cloud environment's *private IP addresses*, you can specify an entire network range with confidence that you can't access hosts outside your cloud. To address ranges within the cloud's private network, you'll need either a scanner within that range or a remote connection, such as a VPN.

If you scan your cloud services' *public-facing addresses*, you'll need to address your hosts individually rather than by network range to ensure you don't accidentally start an unauthorized scan (in other words, an attack) of another organization's hosts. Even though you can more safely scan hosts via their internal addresses, external-facing scans in concert with internal scans help you understand your public-facing vulnerabilities. A vulnerability that only exists on internal-facing services might be less severe than the same vulnerability on a port that's open to the internet at large. Getting both views of your environment will give you a better understanding of your overall security posture.

If you perform internal and external scans, you'll have to make some decisions about the structure of your host data in your database. The scanning and reporting scripts in this book uniquely identify each host by IP address. If a host has more than one IP address, you'll need to account for this by choosing a different unique host identifier. Or, you can treat the external and internal views on the same cloud system as separate hosts. Whichever you choose, adjust your scripts and database to compensate.

Other Implementation Considerations

You'll need a complete understanding of your cloud environment(s) for complete scanning and reporting coverage. Consider the following questions: is your cloud environment largely located in the same place, or is it distributed? Do you have multiple private cloud environments or just one? Is there internal segmentation providing limited access into certain subnets? This section discusses aspects of your cloud environment that you'll need to keep in mind while designing your cloud-scanning system.

Cloud Environment Distribution

Many organizations have multiple cloud environments, possibly spread across several cloud providers, such as Amazon, Google, or Microsoft. Even a "simple" multi-cloud environment might easily include a development cloud environment, a testing cloud, the production cloud environment where the actual business-critical services reside, and a management cloud that controls access into the other three.

Underlying peering connections might link the disparate clouds, or they might be restricted to communication over the public internet. In multiple cloud environments hosted by a single cloud provider, a peering arrangement might allow services in one environment to communicate with another directly. Place your scanners where it's easiest to ensure full coverage of your multiple cloud environments.

Virtual Machines and Services

You can think of a cloud environment much like a traditional data center except all the physical services are replaced by virtual machines. But cloud environments are a lot more flexible. All the major cloud vendors now provide, in addition to custom virtual machines, *software-as-a-service (SaaS)*. In SaaS environments, you can register, say, a PostgreSQL server without having to think about or even be aware of the underlying OS and support software. For the purposes of your business and vulnerability management system, the only thing that exists is PostgreSQL, and the cloud provider handles the patching, configuration, and underlying OS.

Many modern cloud environments have a blend of full virtual machines, SaaS services, and a containerized environment, which I discuss in the next section. You'll need to be aware of this blend and choose your networking settings accordingly to ensure that your scanner can access all open ports across your environment.

Containerized Services

Organizations are increasingly turning to container-based deployments for new services, using systems like Docker and Kubernetes. A full introduction to containers is beyond the scope of this book, but you can think of them as extremely stripped-down virtual machines that expose only specific ports/services to the outside world, if at all. In some cases, especially in Kubernetes environments, you might have multiple microservices that speak only to each other and to the Kubernetes management system; hence, they're nearly invisible from an external scanner's perspective.

Like SaaS systems, containerized environments raise questions of just how much responsibility you have for vulnerability awareness and scanning in these environments. Unlike with SaaS, your organization is still responsible for the containerized environment, even if the environment only externally exposes a very limited set of services. So you need to ensure that the individual containers are not running vulnerable or outdated services. The vulnerability management system we've built in this book is not well suited to managing a containerized environment, but the principles you've learned will serve you well in designing policies to keep these deployments fully up-to-date.

Scanner Access Requirements

To accurately catalog vulnerabilities in your cloud environment, your scanner needs network access to all of your virtual machines and services in the cloud environment. In networking terms, this means that the scanner, wherever it's located, must be allowed to connect to its target IP range on the full range of TCP ports. But what about a SaaS PostgreSQL database? Which ports need to be opened to ensure the scanner can get as much information as possible about that system?

You could allow the scanner access to all ports, 0 through 65535. But considering the database only provides access on port 5432, you might allow the scanner access to only that port on the SaaS host system to save time and effort. On the other hand, what if you don't entirely trust your cloud provider to expose only the PostgreSQL service? The best way to find out what other services are open might be via a comprehensive port scan.

Summary

In this chapter, you learned ways to expand your vulnerability management system. You built a simple REST API to remotely query the vulnerability database to integrate this system with other security or orchestration tools in your environment. You considered the pros and cons of automated exploitation of known vulnerabilities in your environment. You also considered how to extend your vulnerability management capability into the cloud.

Security is always a process, never a goal, and your vulnerability management system is no different. In the next (and final) chapter, we'll look back at what you've accomplished. Then we'll explore some of the topics you might want to tackle next. For example, you might want to investigate the vulnerability management implications of coming trends, such as the zero-trust network, or you may someday want to find commercial replacements for some of your homebrew tools.

15

CONCLUSION

Throughout this guide, you've built a complete vulnerability management system from scratch using freely available tools and some Python "glue." In the process, I hope you've transformed your organization's approach to vulnerability management. But before you close this book, let's look back at where we've been and then look forward to other improvements you might make to your system in the future.

A Look Back

Think about what you wanted to achieve when you first chose this book and started reading. Perhaps you're an IT administrator for a small business who realized that you needed to systematize your patching cycle. Or maybe you're a security analyst who was tasked with formalizing a vulnerability management program in your organization. Most likely, you had a very small budget or no budget at all and had to get creative to source the required hardware for this project. Your goal, even if you hadn't explicitly articulated it, was to gain a comprehensive view of the hosts in your environment and their current vulnerability status.

Designing and Building

When planning and writing this book, my ambitions for the vulnerability management system were twofold:

- Use only freely available, off-the-shelf tools
- Create a system that's comprehensible and extensible

You can decide if we built a comprehensible, extensible system. But it's clear that we built it for no cost beyond the hardware (physical or virtual) to host it. This objective was particularly important to me because vulnerability management is the foundation of a good information security program. In addition, commercial vulnerability management tools are often well outside the budget of organizations with few or no security personnel.

Throughout this book, I've suggested ways to modify the provided scripts and underlying tools to suit your environment and get the best and most actionable data for your organization. Beyond that tweaking, I hope you used the scripting language, database, and other tools that you're most comfortable with. By building and adjusting this system, you now have a powerful tool that you understand intimately enough to improve as your needs change and the vulnerability landscape evolves.

Maintaining the System

Depending on how you configured your vulnerability management system's automation (in *automation.sh*), you might have few to no manual tasks. The system does its work uninterrupted, and every week you receive a series of reports in your inbox or shared folder drawn from fresh scan data. But this doesn't mean you can ignore the system going forward! You need to maintain the system's components and tweak the scan and reporting parameters to sustain and improve the vulnerability intelligence you have to work with.

Although the system as described in this book will automatically update its OS packages, tools, and CVE and exploit data from the internet, keep an eye on these updates. It's unlikely that cve-tools will be unable to update on their own from the NVD repository in the foreseeable future. But you might have to find a replacement for third-party data, such as the Exploit Database-to-CVE mapping. Even if these data sources remain stable, new data sources that weren't available at the time of this writing, or at the time you built the system, could become available. Keep up-to-date on the vulnerability management field and investigate new data feeds to see whether they can improve your vulnerability intelligence.

As your system expands, you might need to expand its hardware as well. If your system is fully virtualized, this might be as simple as assigning further resources to it. But if you're using physical hardware, you might need to roll up your sleeves and perform some physical upgrades. Physical hardware fails and becomes obsolete, so monitor and maintain the vulnerability management system like any other servers in your infrastructure. Don't neglect this maintenance, or you'll risk a system failure just as your CTO asks you about the latest Windows Server zero-day attack that's already being exploited worldwide.

Sooner or later, it might make sense to offload some of your vulnerability management system to the pros: commercial vulnerability management tools and systems. Let's look at how you can bring commercial products into your homegrown ecosystem without losing any of the valuable information that you're already receiving.

Commercial Vulnerability Management Products

Once your vulnerability management process and outcomes have proven successful, your organization might make more funding available. If so, you can start looking at commercial tools to improve your overall system. In this section, we'll consider some of the aspects of replacing part, or all, of your system with commercial tools.

Commercial Scanners

The first step you should take is to research commercial vulnerability scanners and select one to replace OpenVAS. Although OpenVAS is a serviceable tool, commercial tools are updated more regularly. Also, they're much easier to use and include additional features, such as a client-side agent (for example, Tenable's Nessus Agent or the Rapid7 Insight Agent) that can limit or obviate the need for external scanning.

My purpose here isn't to suggest or steer you away from a specific tool. Instead, I'll give you some points to consider when choosing a scanner that meets your needs and that you can insert into your existing vulnerability management system.

Report automation and export
    To build your new scanner into your existing vulnerability management system, you'll need the capability to start scans via your automation script and import the scan results into your database to generate reports. A scanner that generates easily parseable XML or JSON reports will require minimal additional work to fit into your system.

Extensive and documented API
    Generally, commercial scanners provide an API to control the scanner and share scan result data with other tools. The better and more usable the API, the easier it will be to integrate this scanner into your existing system.

Extensible architecture

Many commercial scan tools let you extend the system by adding more scanners to increase coverage. If you choose a scanner that can aggregate results from multiple scanner instances in one central location, you can pull the scan results from that aggregator rather than communicating with multiple scanner instances throughout your network.

Commercial Vulnerability Management Systems

Suppose vulnerability management helps you patch a serious vulnerability that becomes actively exploited a week later or discover an ongoing incursion that your other security tools hadn't yet noticed. Suddenly more money is available—a lot more—to improve the technology underlying your vulnerability management process. Now you can look at full-featured commercial vulnerability management systems.

It might seem as though you've wasted time and effort building a homebrew solution only to replace it with a commercial product, but remember that your system is a means to an end. The goal is to improve the organization's vulnerability posture, and your homebrew system has served its purpose. In addition, by building and maintaining your system, you understand how a vulnerability management system works and can set up and maintain your commercial system as a seasoned expert.

As in the previous section, I won't make any recommendations, but I'll point out a few criteria to help you choose the commercial product that's right for you.

No loss of functionality

The commercial tool should perform the same tasks as your homebrew system or at least integrate with your existing system to share data.

Ability to import existing data

Choose a commercial tool that lets you import data in JSON, XML, or any other open format that you can write a script to generate so you don't lose all the historical vulnerability data you've already gathered.

Ability to export existing data

Don't get locked into a product that won't let you export its data into an open and well-documented format, such as JSON or XML. This is the bare minimum of connectivity you should expect; direct integration with other security tools is even better.

Extensive and documented API

As you learned in "Building a Simple REST API" on page 140, even a simple API can let you share your vulnerability information with other tools. An extensive and well-documented API lets you build customized integrations among your various security tools.

An Incomplete List of Commercial Options

Here's a short, alphabetically ordered list of popular commercial vulnerability scanners and vulnerability management systems to get you started. No recommendation should be implied by a scanner's presence on or absence from this list.

Alert Logic *(multiple products)*

Greenbone Networks GmbH *Greenbone Security Manager*

IBM *QRadar Vulnerability Manager*

Qualys *Vulnerability Management* **and** *Cloud Platform*

Rapid7 *InsightVM*

Tenable *Nessus* **and** *tenable.io*

Tripwire *IP360*

Coming Trends in Information Security

Although the vulnerability management system you've built is well suited to your current organization and network environment, it's worth looking at future trends that might change your information security needs. Consider how cloud, containers, and zero-trust networking will influence how your organization addresses vulnerability management.

Clouds and Containers Revisited

Even today, some organizations, primarily startups and other fast-growing technology companies, have no on-premise infrastructure. All their production systems are in a private cloud. They're managed by an infrastructure orchestration tool, such as Terraform, that dynamically sets up and tears down hosts and services based on current needs. This makes it difficult to determine what systems are currently running, never mind what their vulnerability posture might be. Integrating a vulnerability management system into such an environment will take some thought and likely some cooperation with your *development operations (DevOps)* team.

If you're using an orchestration tool to build and tear down infrastructure, you can build a step into this process that registers (or deregisters) the new host or service with your vulnerability management system. As a result, you'll always have an up-to-date list of hosts and IP addresses that you need to scan. This works well for long-lived virtual hosts, but what about ephemeral hosts that have a life span measured in days or hours rather than weeks or months?

Designating ephemeral hosts as out of scope and delegating their security to the team that builds and maintains them is valid but shortsighted. It's true that there is limited utility to scanning such systems and keeping their results for far longer than these hosts even exist. But the vulnerability management program (that is, you) still has an important role in improving the security posture of ephemeral infrastructure. You can't point to scan results to suggest regular patching of these systems; however, you can insist that updated patches be an integral part of the build process. Either these systems must be fully patched as soon as they're brought online, or, better yet, your organization should create an organization-specific system image that you use as the template for all short-lived systems. Regularly scanning and updating this template is the best way to ensure that short-lived servers are as secure as possible.

Some software tool must exist to create and delete the other systems and ensure that the correct images are used, whether that's Terraform, Kubernetes, Chef, or another automation tool. The build/configuration system is long-lived and is a good target for attacks on ephemeral infrastructure. Use traditional vulnerability scanning and management to secure the build tools.

Organizations that have moved fully to the cloud for their infrastructure or have been cloud-native from the beginning often have their employees connect from arbitrary locations to do their work on their devices rather than company-owned workstations. Such a fully decentralized organization might use a zero-trust networking model, which is the next topic.

Zero-Trust Networking

First described by John Kindervag in 2010, the basic premise of zero-trust networking is simple: don't trust anything without explicit verification. The trust model in traditional network security is based on a network perimeter, allowing devices in some network regions (based on IP address) access to resources and blocking access from all other IP addresses. Zero-trust networking dispenses entirely with the concept of a network perimeter. Devices are authorized individually based on other characteristics set by the network administrator. For example, a system is authorized to connect to resources only if a known user is logged in using multifactor authentication (MFA), the system's onboard antivirus reports that it's clean, and the MAC address is on a whitelist. The goal is to permit only devices that are safe and authorized by metrics other than just an IP address.

Currently, the most prominent zero-trust model is BeyondCorp, a framework developed at Google. Since 2011, Google has been building and using the BeyondCorp model internally and has published several research papers detailing its implementation. To Google Cloud customers, the company provides a zero-trust implementation called "context-aware access," which is modeled on BeyondCorp. Not to be outdone by Google, Microsoft announced a zero-trust framework built around Azure Active Directory. Amazon hasn't publicly released a zero-trust framework in AWS (as this book goes to press), but most of the individual components are available, although some assembly and a third-party identity provider are required.

Zero-trust networking changes the vulnerability management process. Instead of scanning and managing a well-defined set of network segments, you must consider any number of workstations, laptops, and even mobile devices as part of your infrastructure. But how can you regularly scan and remediate vulnerabilities when you don't even know where those devices might be on your network from one day to the next?

The answer is as simple in concept as it is complex in execution: you integrate a vulnerability management metric into the zero-trust authorization criteria. A given host must be clear of major vulnerabilities (or whichever threshold you prefer) to be authorized to connect. This provides motivation not only to ensure your vulnerability data is always completely up-to-date but also to make it an integral part of your overall network security. Building a completely zero-trust networking model takes time and is likely to be an iterative process unless you're building a new network from scratch. But you'll have opportunities to advocate for including vulnerability management in your organization's zero-trust model.

Traditional scanning won't do the trick to get correct and regularly updated vulnerability information for hosts that are constantly moving around and changing IP addresses. You'll need a vulnerability-reporting agent, a small binary that lives on the host and regularly reports back to a central location on the machine's vulnerability state. Because many zero-trust configurations already require an antivirus/antimalware agent on hosts, you might be able to get vulnerability information from the same agent. Given sufficient time and expertise, you could most likely build a homebrew solution that provides similar information, although this is well outside the scope of the current discussion.

In Closing

You've reached the end of this book. I hope you've found the process of building a vulnerability management system as rewarding as I've found the process of writing this book. From day one, I learned new features and details about Python, Mongo, command line tools, and the vulnerability management field. My goal was to transmit much of that knowledge to you.

As with all fields of human endeavor, vulnerability management is not and will never be completely explored. This book is a snapshot of one aspect of vulnerability management. But even between writing and publication, new vulnerabilities, new vulnerability management products, and new ideas about the best ways to catalog and address vulnerabilities have certainly emerged. By working your way through this book and building a customized vulnerability management system, you're in an excellent position to keep abreast of the field as it evolves and even contribute to the state of the art.

Keep in mind that all the scripts in this book are available on GitHub at *https://github.com/magnua/practicalvm/*. If you improve on these scripts or the overall vulnerability management system, please submit a pull request or suggestions. I look forward to seeing all the community suggestions to make this free vulnerability management system even better.

Now go forth and protect your organization's infrastructure!

INDEX

V

vulnerability
    and bugs, 4
    code execution, 4
    command execution, 4
    denial of service (DoS), 4
    information disclosure, 4
    information modification, 4
    prioritization, 15
vulnerability intelligence, 7
vulnerability management
    and risk management, 10–11
    API. *See* application programming
        interface (API)
    cloud environments, 149–152
    life cycle, 5
    mitigation, 9, 32–33
    patching. *See* patching
    recommendations, 9
    remediation, 9
    systemic measures, 33–34
    validating controls, 34–35

vulnerability management system
    cloud environments, 159–160
    commercial products, 158
    components, 49
    database cleanup, 100–101
    hardware requirements, 48
    maintenance, 98–101, 156–157
    Practical Vulnerability
        Management repository,
        xxiv, 161
    prerequisites, 48
    reporting. *See* reporting
    update script, 53–54
    updating, 53–55
    zero-trust networking, 160–161
vulnerability scanner. *See* scanning,
    scanners
vuln-report.py, 111–112, 117, 118

Z

Zenmap, 67
zero-trust networking, 160–161
    BeyondCorp, 160

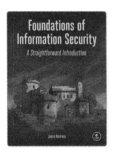

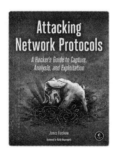

The Electronic Frontier Foundation (EFF) is the leading organization defending civil liberties in the digital world. We defend free speech on the Internet, fight illegal surveillance, promote the rights of innovators to develop new digital technologies, and work to ensure that the rights and freedoms we enjoy are enhanced — rather than eroded — as our use of technology grows.

EFF.ORG

ELECTRONIC FRONTIER FOUNDATION

Protecting Rights and Promoting Freedom on the Electronic Frontier